*The Platypus
and the
Mermaid*

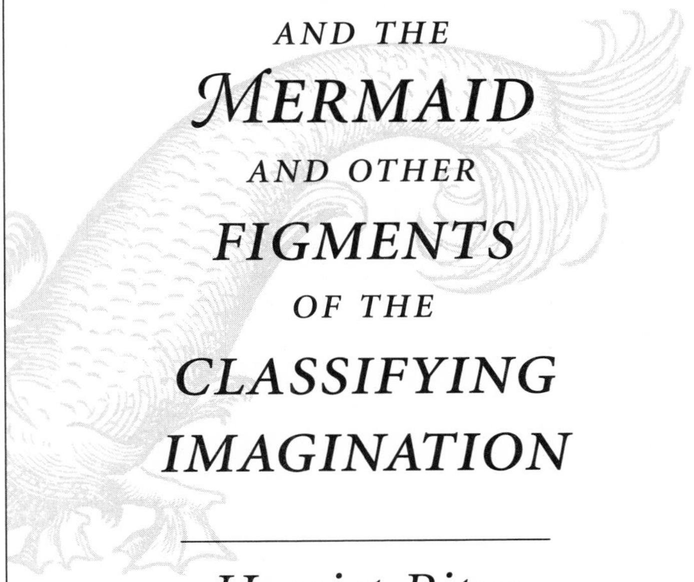

THE

PLATYPUS

AND THE

MERMAID

AND OTHER

FIGMENTS

OF THE

CLASSIFYING
IMAGINATION

Harriet Ritvo

HARVARD UNIVERSITY PRESS

Cambridge, Massachusetts, and London, England

Designed by Marianne Perlak

Library of Congress Cataloging-in-Publication Data

Ritvo, Harriet.
The platypus and the mermaid, and other figments of the
classifying imagination / Harriet Ritvo.
p.   cm.
Includes bibliographical references and index.
ISBN 0-674-67357-3 (alk. paper)
1. Zoology—Great Britain—Classification—History—18th century.
2. Zoology—Great Britain—Classification—History—19th century.
3. Natural history—Great Britain—History—18th century.
4. Natural history—Great Britain—History—19th century.
5. Popular culture—Great Britain—History—18th century.
6. Popular culture—Great Britain—History—19th century.
I. Title.
QL351.R57   1997
590′.1′2—dc21
97-405

For Fran and in memory of
Arthur, Nana, and Herb

Contents

Considered as an Author, Herr Teufelsdröckh has one
scarcely pardonable fault, doubtless his worst:
an almost total want of arrangement.

THOMAS CARLYLE, *Sartor Resartus*

Original caption:

ZOOLOGY.

Railway Porter (to Old Lady travelling with a Menagerie of Pets).

"'Station Master say, Mum, as Cats is 'Dogs,' and Rabbits is 'Dogs,' and so's Parrots; but this ere 'Tortis' is a Insect, so there ain't no charge for it!"

Introduction

A *Punch* cartoon of 1869 featured a railway porter astonishing a prospective traveler with the news that "cats is 'dogs,' and rabbits is 'dogs,' and so's parrots; but this ere 'tortis' is a insect." For Victorian readers, as for their successors, the joke depended on identification with the hapless recipient of this disinformation, with whom they shared the smug certainty that parrots were not, in fact, dogs, nor turtles insects. Their confidence rested, ultimately, on the assertions of a group of experts who claimed the natural world as their intellectual province. But if zoological analysis incontrovertibly demonstrated that turtles and insects belonged in different categories— that the skeletons and circulatory systems of the Testudinae firmly allied them with other Reptilia, such as snakes and lizards, rather than with Insecta, such as wasps and roaches, which were not even vertebrates—the story did not end there. Possibly the railway regulations recalled the earlier vernacular sense of *reptile,* which, more directly reflecting its Latin parent, referred generally to creeping or crawling animals. And the quotation marks that surrounded several of the important categorical terms in the caption suggested that the speaker was aware of the theoretical weakness of his position, as well as its practical invulnerability. But whatever its antecedents and its ironies, this alternative classification did not depend on any of them for its persuasiveness. After all, even the most self-consciously enlightened rail customers—the experts themselves—had to adopt, at least temporarily, the taxonomic perspective embedded in the schedule of rail freight charges, if they wished to transport their subjects and specimens.

The dichotomy on which this cartoon depended was just the tip of the iceberg. Railway bureaucracies were not the only British interest groups to develop systems of classification that reflected their particular needs and experience. Among many others, butchers and artists, farmers and showmen all deployed distinctive taxonomies in their work, although they seldom bothered to articulate them theoretically. Scientific systematizing was similarly polymorphic. By the mid-Victorian period, zoological and botanical classification had relinquished the cutting-edge status that they had held through much of the eighteenth century, but they continued to provoke controversy. Although specialists agreed among themselves that, in general, their classification was superior to any alternatives, there was much on which they differed, ranging from large theoretical issues to the proper location and naming of individual species.

Indeed, the claims of experts—whether they called themselves naturalists, comparative anatomists, or zoologists—that systematic classification represented their appropriation and mastery of the animal kingdom were thus liable to be contradicted from without as well as undermined from within. But if the experts resisted granting recognition to competing claimants of the zoological territory they had staked out, they tacitly acknowledged the objections of various laymen in many ways. They even quietly incorporated vernacular categories into their classificatory schemes, especially with regard to mammals, the creatures most important to people and most like them. This consistently inconsistent practice illuminates both the nature of scientific enterprise during the period and the relation of science to the larger culture. In particular, the determination to ignore or deny genuine sources of influence may have further implications, for this struthious habit has continued to characterize both the scientific community and what is now referred to as the educated general public, with consequences for the design of curricula and the funding of research, among other things.

As anthropologists have repeatedly pointed out, the classification of animals, like that of any group of significant objects, is apt to tell as much about the classifiers as about the classified. In a large and complex society, such as that of eighteenth- and nineteenth-century Britain, animals performed many different functions and stood (or flew or swam) in relation to many different groups of people. Each of the ways that people imagined, discussed, and treated animals inevitably implied some taxonomic structure.* And the categorization of animals reflected the rankings of people both figuratively and literally, as analogy and as continuation. That is,

---

*I assume that people are animals too, but I will nevertheless use the conventional "humans and animals" formula, since it more closely reflects the views of my subjects.

depending on the circumstances, people represented themselves as being like animals, or as actually being animals. For example, worries about the concupiscence of human females structured the theory and practice of animal breeding, and the emergence of racially based nationalism conditioned discussions of species, variety, and breed in animals. More generally, the drawing of boundaries that lay at the heart of any taxonomy resonated strongly with widespread concerns about the stability of established social categories in the face of constant pressure at home and abroad. The relationships between alternative systems were similarly various: sometimes they seemed completely independent of each other, even within the mind or practice of a single person; sometimes assumptions or preoccupations overlapped; sometimes open disagreement or hostility emerged, reflecting fissures in the social fabric or contested cultural territory.

The organization of this book, like its title, attempts to represent the range of these taxonomic practices and also to situate the technical classification of animals in eighteenth- and nineteenth-century Britain within a larger context. The structure is topical, rather than chronological. The chapters become, to put it one way, increasingly vernacular in their focus. The book begins with a consideration of explicitly zoological classification, then proceeds to the taxonomies expressed by its associated nomenclature, by ideas about hybridity and cross-breeding, by the display and interpretation of monstrosities and monsters, and finally by the hunting and eating of animals. Much of the exposition is based on reading widely varied sources as if they constituted a single many-stranded discourse—a Babel with no dominant voice, or a marketplace of ideas and information in which consumers exercised free and even willful choice. To an inevitably limited extent, I have tried to reconstruct the experience of those who were exposed to this voluminous stream of material at first hand, unprotected by either the clarifications or the distortions of hindsight.

As this method has offered fresh perspectives, it has also, as is always the case, required corresponding deemphases. For example, while it foregrounds historical actors and relationships that have often been neglected, it can also conflate groups of participants and blur the social location of individual contributors. Understanding British naturalists and their institutions as constituents of a distinctive national culture means neglecting their significant international dimension, as well as the extent to which parallel developments occurred in other metropolitan western cultures. And stressing the often unacknowledged persistence over time of many taxonomic notions has meant that some celebrated advances, particularly those relating to evolutionary theory, have faded into the background. Fortunately, in recent years, all these areas have been the subjects of distinguished scholarly investigation, from which I have benefitted greatly in my complementary attempt to evoke the elaborate polyphony that formed their context.

*The Platypus
and the
Mermaid*

1

The Point of Order

ONE FAIR WINDY DAY in late June of 1770, as the *Endeavour* lay grounded off the northeastern coast of Australia, those of the crew not engaged in repairing the vessel observed an intriguing animal. In the words of Joseph Banks, no ordinary ship's naturalist but one of the richest men in Britain, it was "as large as a grey hound, of a mouse colour and very swift." James Cook, the ship's captain, added that it had "a long tail which it carried like a grey hound, in short I should have taken it for a wild dog, but for its walking or runing in which it jumped like a Hare or a dear."[1] Over the next weeks they caught occasional glimpses of the unknown creature, which continued to surprise them—most strikingly when they realized that "instead of going upon all fours this animal went only upon two legs, making vast bounds."[2] Finally, in mid-July, Banks happily recorded that the *Endeavour*'s second lieutenant had shot a specimen of "the animal that had so long been the subject of our speculations" and that on close investigation it proved to bear "not the least resemblance" to any animal he had ever seen. The shortness of its forelimbs and the length of its hindlimbs appeared especially remarkable. But if the creature was thus difficult to place within the animal kingdom—its oddities strained the resources of the English language as well as those of scientific taxonomy, so that Banks had to borrow the term *kangaroo* from a local dialect—from a more functional perspective it was easier to pigeonhole. Banks and Cook concurred in proclaiming the otherwise unclassifiable new discovery "excellent food."[3]

The kangaroo skins and bones that the explorers brought back to London soon initiated what was to be a sustained and committed relationship between these unusual animals and Britons with zoological inclinations, a group by no means limited to serious naturalists. Thus in 1790 Alexander Weir, the proprietor of a natural history museum in Edinburgh, enticed potential subscribers by announcing his acquisition of "that extraordinary Quadruped called THE CUNQUROO . . . being the first that ever was brought to Britain."[4] The breadth of the kangaroo's appeal became particularly noticeable when living specimens began to arrive. George Stubbs painted a portrait of what was somewhat ambiguously called "Captain

One of the surprising kangaroos, with special attention to locomotion.

Cook's Kangaroo," and George III installed a few in his menagerie.[5] The immigrants adapted enthusiastically to their new homeland. By the end of the eighteenth century they had become sufficiently identified with Great Britain that an agent of the Florentine government automatically turned in that direction when seeking a *"Macropus giganteus"* for the Grand Duke.[6] In 1804, on the strength of the longevity and fecundity of the royal kangaroos, William Bingley pronounced them "in a great degree naturalized in England," which was likely to "render this most elegant animal a permanent acquisition to our country."[7] Kangaroos were a staple of Victorian public menageries and private parks; by mid-century the Regent's Park Zoo in London regularly offered surplus stock for sale and the Earl of Derby had bred five different kangaroo species at Knowsley Park.[8] By 1878, as William Henry Flower announced in his presidential address to the Zoology and Botany Section of the British Association for the Advancement of Science, it had become "difficult . . . to imagine a world without kangaroos."[9]

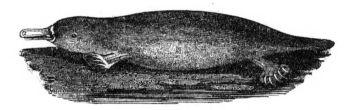

A cryptic early platypus.

The emergence of the still more enigmatic platypus into British consciousness cannot be dated so precisely. In an account of New South Wales published in 1802, David Collins mentioned that he had observed "an amphibious animal, of the mole species," five years previously, but the first specimens did not arrive in Britain until a year or two after this sighting, when they quickly attracted the attention of puzzled naturalists.[10] In his popular handbook *A General History of Quadrupeds*, Thomas Bewick described the platypus that he had examined at a meeting of the Literary and Philosophical Society of Newcastle upon Tyne as "an animal *sui generis;* it appears to possess a three fold nature, that of a fish, a bird, and a quadruped, and is related to nothing that we have hitherto seen"; while the more learned George Shaw of the British Museum, who published the first scientific description of the platypus in 1799, judged it "of all the Mammalia yet known . . . the most extraordinary in its conformation; exhibiting the perfect resemblance of the beak of a Duck engrafted on the head of a quadruped."[11] Indeed, so astonishing did his first encounter with platypus remains seem to Shaw that he found it "impossible not to entertain some doubts as to the genuine nature of the animal, and to surmise that there might have been practised some arts of deception in its structure." Ultimately, his suspicions were laid to rest by the arrival of additional specimens identical to their predecessor in the most troublesome respects, but the scissor marks he left on the original specimen, where he thought that an unscrupulous taxidermist might have attached the beak, bore lasting witness to his initial skepticism.[12]

The taxonomic debates sparked by these puzzling animals continued for decades in scientific circles; as the president of the Royal Physical Society noted in 1831, "no animal has ever excited the curiosity of naturalists more than the platypus."[13] Late Victorian donors were still sending platypuses to the Oxford University Museum in bunches: five in 1893, seven in 1894.[14] Nonspecialists were also fascinated by the platypus's anomalous nature. This wider audience did not offer any alternative mode of classification, either implicit or explicit, even an anthropocentric one based on utility. Perhaps this was because the platypus was much smaller than the kangaroo, as well as more difficult to catch and to maintain in captivity; perhaps because the European animal it suggested to its discoverers was the lowly mole rather than the fleet and handsome greyhound. Indeed, so unprepossessing was the platypus in its native streams that one of its most affectionate early chroniclers—a naturalist who recorded with genuine sorrow the deaths of some young animals he had cared for—characterized its appearance as "sordid and far from attractive . . . resembling rather a lump of dirty weeds than any production of the animal kingdom."[15] Further, one of its most unsettling anatomical features, the cloaca, "which serves both for the functions of reproduction and for the ordinary evacuations," was, in the words of a naturalist concerned with public sensitivities, "highly curious, but not well adapted for popular details."[16]

The appeal of the platypus to the general public, as to naturalists, seemed rather to depend on its weirdness than on any more positive charm or utility. As Charles Darwin wrote of a successful platypus-hunting expedition in New South Wales, "I consider it a great feat to be in at the death of so wonderful an animal."[17] Its stuffed remains and its image figured in non-specialist contexts much more frequently than did those of any other exotic animal of similarly insignificant size and aspect. At times it could represent the odd preoccupations of scientists, as in a satiric *Punch* depiction of "The Meeting of the Zoological Society," where, formally labeled as "*Ornithorhynchus*," it occupied the foreground of the table around which the learned gentlemen gathered.[18] More often, however, the hard-to-place creature appeared firmly, if somewhat paradoxically, integrated into the familiar domestic scene. In 1851 the Natural History and Antiquarian Society of Penzance figuratively placed a platypus among the overwhelmingly local fauna on display in its museum; and fifteen years later some members of the Acclimatisation Society of Great Britain more literally included the platypus among the foreign species that might profitably be naturalized at home.[19]

The Mammalian Other

Fascinating though the Australian animals indisputably were, their physical endowments might not have been sufficient by themselves to attract the persistent attention of a wide range of audiences. After all, the kangaroo and the platypus appeared on the British scene after several centuries of vigorous global exploration, one recurrent result of which had been the discovery of such previously unsuspected animals as the armadillo and the sloth. Many of these novelties had been transported home for display, dead or alive, along with other creatures, such as the Indian nilgai or blue bull, that had been for centuries only bookish rumors.[20] No matter how curious or spectacular, however, most of these creatures, like the giraffes and hippopotami that followed them in the nineteenth century, were in effect nine-days'-wonders, enjoying a brief rush of celebrity and then dwindling into routine menagerie and museum displays and conventional encyclope-

A Victorian platypus, naturalized among creatures of British woods and fields.

dia entries. So predictable was this boom-and-bust cycle, and so quantifiable in its effect on gate receipts, that the official marketing policy of the Regent's Park Zoo stipulated that the public be provided with a steady diet of amazing novelties.[21] Both naturalists and zoological idlers might soon have tired even of duck bills and giant feet. What guaranteed the continuing appeal of these animals was the fact that their oddity was not confined to the merely physical but extended to the level of theory or system. Unlike the giraffe, for example, which, although at least equally "singular in its structure" at first glance, could upon closer inspection confidently be assigned "a distinct genus" in the vicinity of the deer and the antelope, to which it was "nearly allied," the kangaroo and the platypus did not seem to be nearly allied to anything.[22]

Further, both kangaroo and platypus were representative rather than idiosyncratic anomalies. That is, each belonged to (and figured as a synecdoche for) a group of animals characterized by at least some of the features that made it seem so strange—marsupials and monotremes, respectively, as they were known to nineteenth-century zoology. The indigenous mammals of the southern continent seemed to have been designed according to a plan different from those that shaped the animals of the rest of the globe. Almost all of them turned out, like the kangaroo, to have pouches, and many shared its peculiar conformation of foot and leg. Among the few nonmarsupials, the echidna, which lacked a birdlike bill and webbed feet, made a less

The echidna was less flamboyant than the platypus, but equally anomalous.

eccentric impression than the platypus. It had received its first scientific description almost a decade earlier, also from George Shaw, without provoking any special furor. But when dissected it similarly revealed organs of reproduction and excretion that, in the view of the anatomist Everard Home, gave "this new tribe a resemblance in some respects to birds, in others to the Amphibia."[23]

In the decades that followed the discovery of the platypus, many naturalists echoed Home's (and Bewick's) inclination to fudge about its relationship to other mammals, as they were beginning to be known among specialists, or quadrupeds, as they continued to be somewhat misleadingly termed by the old-fashioned and the technophobic. For example, Busick Harwood, the professor of anatomy at Cambridge, emphasized the boundary-blurring effect of its multiple affinities in the syllabus for the natural history course he offered in 1812, promising to devote special attention to "the astonishing union of the characteristic distinctions of all the Classes in that extraordinary animal."[24] Some naturalists went so far as to deny the platypus the mammalian status to which its furry (if coarse) pelt seemed to entitle it. After an elaborate technical critique of Home's dissection, John Thomson concluded that "from the want of mammae, and from the structure of the sexual organs, the naturalist surely cannot, with any degree of propriety, arrange this animal with the Mammalia; and very few will be hardy enough . . . to think of arranging it with Birds or Fishes. The only possible class that remains, is the Amphibia."[25] Naturalists who tried to wedge these animals into established mammalian subcategories reached conclusions that were no more satisfactory, whether the odd taxonomic bedfellows they proposed were edentates (toothless animals like sloths and anteaters) as William MacLeay suggested, or palmates (animals with webbed feet) as William Coulson claimed in the introduction to a comparative anatomy manual.[26] Thus it was not surprising that after a few years the specific designation *anatinus,* which Shaw, exercising his right as the first describer of the platypus, had bestowed on it, was gradually supplanted by the more expressive *paradoxus.*[27] Perhaps it was not entirely a printer's error that caused the *Zoological Miscellany* of 1815 to list the platypus and the echidna under the rubric "Monstremata."[28]

Kangaroos and their kin presented analogous if subtler classificatory conundrums. At first, those who recognized marsupials as a coherent group tried to assign them a place within one of the established mammalian orders. But the fit was always awkward, whether they appeared with the monkeys, as in the display cases of Bullock's Museum, because at least some of them seemed to have opposable thumbs, or with the rodents, with which

Didelphis ursina.

Didelphis cynocephala

Marsupials drawn and named to resemble carnivores: *Didelphis ursina*
and *Didelphis cynocephala*

they seemed to share some hab
carnivores because of superficial s
that many marsupials, including
inevitably made this placement so
Quadrupeds Thomas Pennant locate
bears and the weasels, while confessi
was with each other.[29] Not everyone,
pouch as taxonomically definitive. Son
peculiar structure" that seemed to bind
to be less significant than their "many s
in which case, for example, opossums m
and kangaroos with the ruminants that s
at the opposite extreme were a few whe
should be considered mammals at all, a s
the announcement of one Richard Q. Couch that he had examined several
living "marsupial fishes," caught off the Cornish coast.[31] A zoological primer
described the marsupial condition as a taxonomic oxymoron: "the young
grow like buds, which is a character of the very lowest tribes of animated
beings; and then draw milk from the teats of their mothers, which is a
character of the very highest, and of man himself."[32]

Despite this interpretive polyphony, a durable consensus about the nature
of marsupials and monotremes and their proper position within the animal
kingdom emerged among scientists in the course of the nineteenth century.
By 1896, for example, Richard Lydekker, a prolific zoological author and
distinguished curator at the British Museum (Natural History), could happily reflect that "in few, if in any group of Mammals, is our knowledge of
species and genera in such a satisfactory condition as is the case with the
Marsupials and Monotremes."[33] This confident formulation assumed that
both groups belonged in the class Mammalia and that each displayed
internal coherence, based ultimately on distinctive reproductive methods
and apparatus: the marsupial pouch and the monotreme egg.[34] Zoologists
might still debate about the relative positions of these larger units—whether
the mammals should be divided into three structurally equal subclasses (the
third and by far the largest, also defined reproductively, consisting of placental mammals), or whether the extreme divergence and possible independent descent of the egg-layers should be signaled by some more marked
taxonomic segregation, or whether, on the other hand, the monotremes
should be "degraded into a mere subdivision" of the marsupials—but such
disagreements seemed mild, even quibbles, in comparison with earlier differences of opinion.[35] Occasionally, the discovery of an apparently ambigu-

THE PLATYPU

10

ous new species might
Such alarms, howev
These sporadi
to consensu
because
in re

momentarily threaten to reopen the old debates.
er, invariably turned out to be false.[36]
squabbles served nevertheless as reminders that the road
had been neither straight nor smooth. This was at least in part
the stakes involved in its achievement had been so high. Although
rospect the consensus could be (and ordinarily was) seen to represent
the steady and rational interplay of accumulating data and evolving theory,
its persuasiveness and authority had been less clear to earlier naturalists
who encountered it in its incipient stages. Alternative understandings of the
anatomy, behavior, and geographical distribution of marsupials and
monotremes might have seemed equally plausible to them, and broader
allegiances also helped to determine which interpretation they favored. Thus
the classificatory difficulties posed by the aberrant Australian fauna turned
out to have multiple causes and complicated effects. Inevitably, as was the
case with any taxonomic anomaly, even one so readily resolvable as the
long-necked giraffe, these difficulties emphasized the unusual nature of the
animals themselves. But the implications of the less tractable marsupial and
monotreme anomalies also ran in the other direction. Since anomaly is by
definition comparative, their violation of expectations reciprocally called
into question both the zoological assumptions current before their advent
and the systems in which those assumptions were embedded.

The strains placed on the enterprise of classification by marsupials and
monotremes differed significantly from the more ordinary or predictable
problems faced by taxonomists, which could also be severe. In a period
when British commercial and military expeditions regularly returned bearing zoological and botanical specimens from a vast range of territories
previously inaccessible to the scrutiny of western science, the discovery,
naming, and classification of new species was a routine feature—indeed the
staple employment—of natural history. The roster of known animals, which
had begun to increase rapidly by the end of the fifteenth century, continued
to snowball. Between John Ray's late-seventeenth-century catalogue of
quadrupeds and that of Linnaeus half a century later, the number of species
had doubled from 150 to 300; by the late nineteenth century an average of
more than one thousand new genera of animals of all kinds were being
described each year—"a simply appalling number," in the view of a commentator for *Nature*.[37] If an energetic collector was on board, even a single
extended voyage could significantly increase the taxonomic burdens of
naturalists. Captain Cook's travels in the south seas produced hundreds of
previously undescribed bird species and approximately fifty to one hundred
new mammals, not to speak of much more numerous fish, insects, and other

invertebrates. That the numbers could not be specified with greater precision—and that the collections themselves were rapidly broken up—testified to the fact that zoological stay-at-homes were often unequal to the challenge presented by such sudden riches. In the 1830s Charles Darwin had analogous difficulties appropriately disposing of the much larger collections he had amassed during his five years on the *Beagle*.[38]

Bottlenecks in the processing of novelties less potentially unnerving than kangaroos and platypuses reflected failures of application (usually insufficient or unwilling manpower) rather than systematic lapses. Indeed, as the building up of this flood was one inspiration for the intense attention to classification that characterized eighteenth- and nineteenth-century natural history, so the ability to accommodate it, at least in principle, was one of the standards by which taxonomic systems were judged. A system that was too rigid to make places for new discoveries, or too limited by the state of knowledge existing when it had been devised to tolerate occasional realignments, was a system whose time had come and gone. Writing in the popular Naturalist's Library series in 1846, Charles Hamilton Smith announced that "by the progress of science [that is, the constant discovery of new species], the labours of Linnaeus have . . . in great measure, become confused and inapplicable. This was the natural result in a science based entirely on facts."[39] But the challenge posed by the Australian fauna was not merely quantitative—although certainly there turned out to be a lot of marsupials, at least before they began to succumb to "the war of extermination recklessly waged against them" by British colonists.[40] These unusual animals called into question systematic flexibility at a different level, undermining the very categories that could not be stretched to accommodate them, as well as the principles on which those categories were based, and compromising the authority of the experts who endorsed and applied them.

After all, the category principally cast into question by antipodal eccentricities—that of quadruped or mammal—was among the oldest and most stable in the zoological canon, and it included the creatures of greatest economic, intellectual, and emotional significance to human beings, as well as, in the view of at least some naturalists, humans themselves. Aristotle had identified those creatures as viviparous quadrupeds, a varied group whose members shared, in addition to live birth and four feet, blood, hair, and a terrestrial mode of existence.[41] Having passed through the intervening millennia relatively unscathed, Aristotle's category was adopted by seventeenth-century British naturalists with only minor alterations. For example, in his *Synopsis Quadrupedum* of 1693, the pioneering systematist John Ray suggested alternative subdivisions for it while maintaining the same external boundaries.[42] Ray's successors similarly accepted this definition of quadrupeds as the basis for their own research, although they might wrangle about the underlying principle of Aristotle's taxonomy—whether, for example, it was based on circulation or locomotion[43]—or use insights newly provided by comparative anatomy to buttress old assertions that the category "quadrupeds" could nevertheless encompass warm-blooded unfeathered creatures with no feet (whales) and with two feet and two wings (bats). But the arguments for including these creatures within the quadrupedal realm, etymological considerations to the contrary notwithstanding, were precisely the arguments that made the kangaroo and, especially, the platypus problematic.

The fact that these animals juxtaposed incontestably mammalian characteristics like hair and warm blood with others previously identified only with birds and reptiles forced naturalists to consider whether some quadrupeds were intrinsically more mammalian than others. And the systematic oddity of the Australian fauna was, in a sense, contagious. Redrafting the boundaries of a previously well-defined category was not necessarily a matter of simple expansion; new proximity to external classes potentially shifted all internal relationships too. Even the language with which naturalists described the unlooked-for marsupial and monotreme recombinations continued to emphasize bridging and convergence, long after the initial shock had faded. Soon after the first platypuses began to reach Europe, Banks received a letter about "a very curious animal from New South Wales, which connects the Classes of Birds and Quadrupeds."[44] The author of an early Victorian anatomy manual asserted that although birds and quadrupeds "might appear . . . remotely separated . . . the MONOTREMATA . . . form a link between these two great classes."[45] Darwin wrote of marsupials and monotremes that "I cannot see any objection to considering them as links" between mammals, birds, and reptiles.[46] Decades later, a guidebook instructed visitors to the Ipswich Museum that marsupials were "more or less connected with Birds," and that monotremes "bringing up the rear of the Mammalia . . . still more directly connected them with Birds."[47]

Previously, there had been little need to disassemble the package of mammalian attributes or to organize them into a graduated hierarchy, because they had not presented themselves separately in nature. That is, all creatures designated as quadrupeds seemed to possess them all. The only assumption of priority was implicit in the term *quadrupeds* itself. And while zoological authors were not much troubled by the apparent inconsistency of excluding lizards and salamanders—the distinction between viviparous and oviparous quadrupeds descended from Aristotle—the power of nomenclature was sufficient that they did ordinarily feel compelled to explain why bats and marine mammals were, equally traditionally, included. Thomas Bewick carefully informed his readers that, despite its watery habitat and lack of feet, the walrus was "nevertheless classed by naturalists under the denomination of quadrupeds" and that the bat was allied to birds "by the faculty of flying only," but to four-footed animals "both by its external and internal structure."[48] Indeed, the perceived need for such explanations was a primary motive for the late-eighteenth-century replacement of *quadrupeds* by *mammalia,* a term that highlighted the mammae that seemed so notably absent in monotremes.[49]

This replacement was no mere matter of idle wordplay. On the contrary, the emphasis on mode of locomotion as the primary division among vertebrates may have precluded a much earlier suspicion of the taxonomic problems posed by marsupials—or, to put it another way, a much earlier recognition that these animals shared many distinctive and potentially significant characteristics. Despite the professions of astonishment that accompanied their discovery, kangaroos were not the first marsupials to meet the bemused European gaze, nor even the first to attract signficant attention. The opossum was among the most widely admired and discussed of new world animals, as well as one of the earliest to be noticed and displayed. A living specimen from South America had traveled to Spain within a decade of Columbus's first voyage. Naturalists of the sixteenth and seventeenth centuries were fascinated by the pouch and by the fact that the opossum seemed to be composed of parts from a variety of other animals, including the fox, the bat, and the ape, but they leaped to no systematic conclusions. Later in the seventeenth century, several travelers in the East Indies reported chance encounters with creatures that could be retrospectively identified as wallabies (small kangaroos), without causing any stir at all.[50] Certainly no one suggested a link between them and the opossum. When the anatomist Edward Tyson dissected a male (and therefore pouchless) opossum in 1699, he found that, unlike the dogs and weasels with which it was most frequently classed, it lacked a penis bone. This troublesome discovery inspired him to sketch a new classification of land animals, but his redrawn system still focused on locomotive rather than reproductive organs.[51] And if it had had only its pouch to distinguish it, even the kangaroo might have made a less spectacular first impression. After all, Cook and Banks were initially inclined to liken it to a dog. Its unusual mode of locomotion persuaded them that it was a taxonomic anomaly before any suspicions had arisen about its aberrant reproductive or nurturing proclivities.

As the irregularities of marsupial and monotreme anatomy undermined previous taxonomic assumptions and structures, they created a vacuum of zoological authority, if not of power. Old systems of mammalian classification had to be seriously revised, or even discarded, and the ability plausibly to accommodate the Australian fauna became an important criterion for judging potential replacements. Thus marsupial and monotreme classification emerged as a battleground upon which rival systems and rival systematists could engage, sometimes directly addressing the matter at hand, sometimes using it as a stalking horse for issues still larger or deeper. Such confrontations could be civil, as was the correspondence between Charles Darwin and G. R. Waterhouse in 1843, in which, while more or less agreeing about the relationship between monotremes and marsupials, they aired significant differences of opinion about the bases of natural classification.[52] Or play could be very rough, as when, in the 1830s, the brilliant and irascible anatomist Richard Owen entered the fray. As part of his general assault on the evolutionism of radical British followers of Jean-Baptiste Lamarck, Owen denied the intermediate or transitional character of monotremes and categorized them instead with the edentates.[53] So acrimonious was this debate that a decade later Thomas Wakley, the radical editor of the *Lancet,* deplored it as "scientific malversation," shocking evidence of "the cliquism and favouritism which have disgraced" the Royal Society.[54]

In order to defend his position, Owen had to deny that the platypus laid eggs, a denial that implicitly identified another way in which the taxonomy of Australian mammals had become a representative arena for contests between rival authorities. From the beginning, the reproductive tract of the female platypus, which was odd for a mammal but less so for a bird or a lizard, had been a kind of smoking gun, suggesting oviparity. In addition, human denizens of the bush, both British settlers and indigenous Australians, testified repeatedly to the fact that the platypus laid eggs, and they occasionally even showed sample eggs to interested naturalists, who did not, however, feel compelled to believe them. Convincing proof required that a member of the scientific community actually view eggs in intimate conjunction with monotreme reproductive organs. In 1825, Everard Home urged the emigrant W. S. MacLeay to repair this lacuna, emphasizing that "what is principally wanted is the ova." Their continued absence "after so long a period that the colony has been under the British Government" had patriotic as well as scientific implications, being "neither to the Honour of the government, the credit of those who have resided there, or those at home."[55] Despite these exhortations, the feat was not accomplished until 1884, when W. H. Caldwell traveled from Cambridge to Australia for that specific purpose. After several weeks he managed to shoot "an *Ornithorhynchus* whose first egg had been laid; her second egg was in a partially dilated *os uteri*."[56] And although Caldwell immediately telegraphed his momentous news to Montreal, where it was read to the zoologists gathered there for the annual meeting of the British Association for the Advancement of Science, apparently word trickled down slowly to the popular mediators of scientific information. Several years later, while describing a stuffed platypus on display at the Australian pavilion of the Colonial and Indian Exhibition, the *Illustrated London News* smugly reported that "fables were formerly told of this queer creature, as that it laid eggs."[57]

The Invention of Tradition

The kangaroo and the platypus emerged into the scientific and popular limelight toward the conclusion of what was repeatedly hailed, both in its own time and subsequently, as the heroic age of scientific classification.[58] Enthusiasts praised the Herculean labors of the founding systematists, often represented, especially in hindsight, by the single name of Linnaeus. He had made it possible, at least in theory, to assign each animal or plant its own unique position in his comprehensive system, and therefore by implication to offer an objective, rational, and complete analysis of the apparently chaotic and infinitely varied products of nature. The new methodology embodied in Linnaeus's *Systema Naturae,* which was first published in 1735 and repeatedly revised and expanded, redefined the study of the living world, and even seemed to put natural history on a par—or at least in the ballpark—with the more prestigious and mathematically oriented pursuits grouped under the rubric of natural philosophy. Serious naturalists appreciated this realignment and tended to emphasize the philosophic aspects of their discipline. "Without a systematic classification," according to William Turton, the editor of one of the many versions of Linnaeus published in England around the end of the eighteenth century, the student of zoology "wanders in obscurity and uncertainty, and must collect the whole of its habits and peculiarities, before he can ascertain the individual he is examining."[59]

Linnaean classification also served more practical purposes. It conveniently separated people from the other animals: "to study the works of Creation with intelligence is the exclusive privilege of man, and highly exalts his dignity above that of all other animated beings."[60] Besides establishing a natural order and confirming the human position at its head, a well-conceived taxonomic system could help define and dignify the place of both the discipline of natural history and its adherents in the human intellectual order. Benjamin Stillingfleet, another interpreter of Linnaeus, made the ability to give a plant or animal "its true name according to some system" a kind of *sine qua non,* observing sternly that "he who cannot go thus far . . . does not deserve the name of a naturalist."[61] Many of his fellow naturalists viewed the disciplinary stakes as higher than merely the exclusion of the unprepared. Without system, they feared, natural history would be "but a confused, undisciplined crowd of subjects" and naturalists "mere collectors of curiosities and superficial trifles . . . , objects of ridicule rather than respect."[62]

Classification lost its flagship status during the first decades of the nineteenth century, when it was replaced on the cutting edge of zoology by physiology and allied pursuits. By 1834, William Jardine, the editor of the Naturalist's Library, could proclaim that "the age of superstitious reverence for categories . . . has long passed away."[63] Nevertheless, the pioneer taxonomists continued to occupy prominent positions in the zoological pantheon enshrined by Jardine and his collaborators, and the sudden emergence of systematic taxonomy in the previous century continued to symbolize the birth or creation of natural history as a scientific study. The biographical sketches that began each volume of the Naturalist's Library often featured zoologists distinguished primarily for their contributions to Enlightenment systematics, among them Ray, Linnaeus, Pennant, and Francis Willughby (Ray's collaborator).[64] Their heroic rescue of natural history from the dark clutches of superstition was quickly installed among the founding myths of biological science, so that it was ritually recounted even by revisionists profoundly at odds with their esteemed disciplinary fathers on many matters of theory and practice.

The negative side of this coin of celebratory commemoration was at least equally important to self-conscious modernists. As Jardine put it, with the publication of the first edition of Linnaeus's *Systema Naturae,* "the arrangements of the older systematists were almost at once superseded."[65] The rhetoric that celebrated the novelty and objectivity of Enlightenment systematics simultaneously signaled and created a gulf between all succeeding naturalists and previous students of the animal creation. An Oxonian bard politely thanked the

> Guides of my way full many an ancient sage,
> With analytic and synthetic page,

before acknowledging the sources of real wisdom:

> But chief to truths from elder ages hid,
> Linnaeus, Cuvier, Paley, Buckland, Kidd.[66]

More drastically, Enlightenment and post-Enlightenment admirers of John Ray routinely presented him as a prodigy without immediate intellectual ancestors.

Like that of Linnaeus, the standard praise of Ray emphasized the discontinuity that separated his work from preexisting compilations on related topics, placing equal emphasis on Ray's virtues and the defects of the competition. Ray's contemporary Edward Tyson honored him as the father of systematic natural history and comparative anatomy in England, in contrast with earlier pretenders to that distinction, "more Pompous than Instructive," whose method had been "to rake in all from former Authors, without separating the weeds, or sifting the chaff from the Grain."[67] As Thomas Pennant put it several generations later, "so correct was his genius, that we view a systematic arrangement arise even from the Chaos of *Aldrovandus* and *Gesner*."[68] Richard Brookes, who asserted that "no systematical writer has been more happy than he [Ray]," dismissed the writings of Ulisse Aldrovandi as "insupportably tedious and disgusting" and those of Konrad Gesner as "so incomplete as scarce to deserve mentioning."[69] These withering judgments continued to resonate into the nineteenth century. The large and diverse Victorian readership of the Naturalist's Library learned that Gesner's subdivisions were "altogether arbitrary and useless"; that Pliny's "most obvious defect is the want of any thing like system"; and that, in general, "with the exception of Aristotle, neither the philosophers of antiquity, nor those . . . succeeding the revival of learning" studied nature with "that accuracy of observation and reference to organic structure, so necessary for . . . determining . . . the classes, orders, genera, and species."[70]

Although Aristotle's incorporation into the canon of Enlightenment natural history, as well as his extrazoological eminence, exempted him from this wide-ranging and contemptuous dismissal, eighteenth- and nineteenth-century naturalists took care to stress the uniqueness of his position. Most acknowledgments of Aristotle's anachronistic insight were constrained by observations that the real significance of his work had been largely ignored and that little scientific progress had occurred between his time and the beginning of the modern taxonomic era. As Richard Owen put it in 1847, "during the two thousand years which have elapsed since Aristotle wrote and lectured . . . , the ideas of learned men regarding the nature and classification of Mammalia received no improvement, and any change which they underwent was for the worse."[71] And even Aristotle's distinguished reputation could not shield him completely from invidious comparisons with the accomplishments of zoological moderns. As naturalists grew more confident, even this most illustrious precursor began to seem irksome and dispensable. By the latter part of the nineteenth century Aristotelian classification could be disparaged as "scarcely so clearly developed as some of his admirers would lead us to admit" and the very idea that his zoological insights might rank with those of Victorian science as "a very absurd homage to antiquity."[72]

This elaborate and extended dismissal of the past constituted an implicit reification of the zoological present, to which could be attributed, by the logic of negation or opposition, those qualities of rational order, universal application, and expert consensus that the works of outmoded naturalists seemed so signally to lack. Further, the victory of system over ancient and entrenched error appeared to its advocates the harbinger of broader relevance and extended sway. Scientific classification was hailed as both symbol and agent of a larger intellectual triumph, one that could ultimately reverse the traditional relationship between humans and the natural world. Even Linnaeus's humblest admirers shared his sense of the almost godlike possibilities offered by the intellectual mastery of nature. Thus Leonard Chappelow, an East Anglian clergyman who labored for years over a never-published taxonomic epic, sang the Linnaean achievement in biblical terms:

> And thus not only were arranged, and classed,
> The subjects of my vegetable world,
> But every *Beast*, and every *Bird*,
> The amphibious tribes, tenants of the
> Congregating waters of the deep, were
> Summoned all, and all again displayed in order
> To receive from thee their names . . .[73]

Citizens of a prosperous global power like Great Britain easily conflated the metaphorical dominion of knowledge with more practical or literal modes of appropriation. Naturalists in the mother country automatically claimed the right to classify the plants and animals of its growing colonial territories. Although this right of possession was more often assumed than articulated, dispossession could provoke an explicit statement. As Thomas Pennant lamented in a preface of 1784, "this Work was designed as a sketch of the Zoology of *North America*. I thought I had a right to the attempt, at a time I had the honor of calling myself a fellow-subject with that respectable part of our former great empire; but when the fatal and humiliating hour arrived, which deprived *Britain* of power, strength, and glory, . . . I could no longer support my clame of entitling myself its humble Zoologist."[74] And if system was a means of consolidating the intellectual dominion of science over nature, especially in exotic subject territories, at home it also represented the superiority of institutionalized cosmopolitan learning to mere provincial or local lore. As a Swede who had traveled widely in northern and western Europe and whose work had been respectfully received by the naturalists of a variety of non-British linguistic and cultural communities, Linnaeus himself powerfully exemplified the increasing prestige and authority of internationalized expertise.[75]

Sweet Disorder

In the context of this magisterial ascendancy, the confusion attending the discovery and classification of the Australian fauna may have seemed almost as anomalous as the creatures themselves. The extended and frequently cacophonous debate about marsupials and monotremes implied quite a different zoological enterprise—one in which consensus was rare, in which authority was uncertain and fragmented, and in which the very principles behind the construction of taxonomic systems and the assignment of individual species to their niches were vaguely defined and of obscure or questionable provenance. Despite these discrepancies, however, or perhaps because of them, this relatively chaotic portrait of eighteenth-century natural history may have been at least equally recognizable to contemporaries as the confidently progressive interpretation offered by naturalists inclined to celebrate the advent of system. For if the classificatory problems presented by the Australian animals were unusually profound, they did not uniquely suggest an alternative view of the zoological establishment and of the role of taxonomy within it. The systematic position of even such well-known animals as elephants and hyenas remained subject to persistent disagreement.[76] In this respect, at least, the difference between marsupials and monotremes on the one hand and less surprising animals on the other could be seen as a distinction of degree rather than kind; paradoxically, the bizarre attributes that made the kangaroo and the platypus anomalous could thus also make them representative.

Eighteenth-century natural history contained many suggestions that taxonomic triumphalism, however orthodox, was premature. The very documents created to enshrine the Promethean version of the origin and efficacy of systematic classification—the treatises and handbooks published for an ever-increasing and diversifying audience—could simultaneously tell another, rather different story.[77] At a glance, for example, the contrast between the orderly catalogues of the Enlightenment and the miscellanies of preceding ages was both striking and ubiquitous; it was obvious even if unsophisticated natural history primers produced for eighteenth-century children were compared with the most elaborate Renaissance treatises.[78] Edward Topsell's massive *History of Four-Footed Beastes,* which was published in 1607, purported to offer information that was complete, accurate, and up-to-date. But despite the fact that its entries included a good deal of relatively fresh material, it clearly belonged to the bestiary tradition that stretched back through the Middle Ages to antiquity. That is, it was distinguished by both plenitude of detail and paucity of organization; no readily apparent principle determined the order of the entries for different animals or, within the entries, what material should be included—or even what constituted fact. The long discussion of the domestic cat began with the role of cats in ancient Egyptian society, then went on to consider, among other things, which kinds of cats were best, their eyesight, their hunting methods and favorite prey, their capacity to love and hate, the diseases they suffered, the diseases they caused, and the diseases they could cure or alleviate (for example, the fat of a cat might be efficacious against gout), all with reference to such classical authorities as Pliny and Galen.[79] The entry on unicorns was longer, just as detailed, and derived from equally authoritative sources.[80] Under Enlightenment scrutiny, this massive work seemed like a jumble, a promiscuous mix of ancient wisdom and modern observation, which implicitly defined people as passive recipients of random information about the natural world.[81]

Even the superficial appearance of eighteenth-century natural histories was calculated to produce a very different impression. Most authors prefaced their work with both verbal and graphic proclamations of their concern with order. The minimum was an alphabetical table of contents, such as Thomas Bewick used in his very popular but not particularly erudite *General History of Quadrupeds.* Bewick rather apologetically characterized the table as "our disregard of system," even though it did not represent the organization of his entries, which, like those of many naturalists influenced by Buffon, embodied a loose notion of kinds.[82] Naturalists addressing more serious audiences offered more elaborate analyses, including not only contents listed systematically as well as alphabetically but also graphic representations—diagrams, charts, or tables—of the systematic relationships between the major categories of animals.

If most English naturalists of the eighteenth and early nineteenth centuries structured their works to emphasize the theoretical gap that separated them from such unscientific predecessors as Topsell, however, they were not always able to muster convincing evidence of this discontinuity in practice. No matter how systematically or according to what taxonomic method these volumes were organized, descriptive entries about individual species comprised the bulk of their contents. Often, these entries bore strong affinities in both structure and detail to those offered by Topsell and his ilk. For example, Bewick began his entry on the cat with the disclaimer that "to describe an animal so well known, might seem a superfluous task," then went on to discuss its eyesight, its voice, its irritability, its gestation, its lack of affection for humans, its price in medieval Wales, and its relationship to Dick Whittington.[83] Addressing a less popular audience, Pennant noted, among other things, that wild and domestic cats came from the same stock, that Angora cats degenerated in England but were adorned with silver collars in China, that tortoiseshells were black, white, and orange, that cats purred when pleased and washed their faces at the approach of a storm, and that although many people hated them they were much loved by the "Mahometans."[84]

Miscellaneous inclusions of this sort blurred the line separating natural history from other pursuits. And there was little evidence that some of those whose work was taken to have established that line had been conscious of it at all. For example, John Ray apparently found Renaissance cabinets of curiosities, disparaged by subsequent naturalists as the apotheosis of categorical confusion, very helpful in his systematic studies, especially the famous Tradescant collection—also known as "the Ark"—in south London.[85] In fact, this collection itself provided another illustration of the permeability of the chronological and disciplinary line allegedly separating system and chaos. Oddly assorted though it might have been—the first catalogue contained such divisions as "Birds" (including the remains of a dodo), "Strange Fishes," "Outlandish Fruits," and "Warlike Instruments"—the Tradescant collection was ultimately transformed wholesale into the stuff of science, by the simple device of reacquisition and renaming. As part of the Ashmolean Museum, it helped form the core of the university natural history collections at Oxford.[86]

And if the disorderly past turned out to be less remote than many naturalists claimed, the systematic present reciprocally tended to recede. That is, beneath the reiterated consensus about the novelty and value of zoological (and botanical) classification lay a great deal of disagreement and uncertainty about exactly what was being celebrated. Most Enlightenment naturalists joined the chorus of praise for system in the abstract; but their responses to particular systems were apt to be less cohesive. The very icons of classification—the tables and diagrams prefixed and appended to works of Enlightenment zoology to distinguish them from the unstructured productions of previous ages—could simultaneously evidence this lack of unity.

In *A Cabinet of Quadrupeds*, John Church hesitated between two schema, confessing that the "systematic arrangement . . . of Mr. Pennant . . . takes the lead; but for the use of those who may prefer the Linnaean arrangement, it has been added"; following the preface, both systems were displayed in tabular form.[87] Similarly, one of the late-eighteenth-century translators of Buffon included a chart in which Buffon's genera were laid out against those of Pennant.[88] Several decades later the cataloguer of the Ashmolean collection conducted an openminded survey of available taxonomic options for organizing museum displays, before devising one "derived partly from Linnaeus, partly from Cuvier, with additions and improvements." To illustrate his difficulty—an embarrassment of riches—he included a synopsis of the rival arrangements of mammals propounded by Linnaeus, Blumenbach, Cuvier, Illiger, Fleming, and Latreille.[89]

The multiple possibilities demonstrated by such tables also called into question the standard synecdoche by which Linnaeus represented systematists in general. Admiration for Linnaeus was, to be sure, frequently and fulsomely expressed by his British contemporaries. For example, Joseph Banks referred to him as "our Master" in a letter to Thomas Pennant, and by the end of the century, according to a writer on agriculture and natural history, "the system of Linnaeus has obtained such marked approbation . . . as to supersede the necessity of . . . adverting to it."[90] But even Linnaeus's sincerest admirers might qualify their praise. While John Berkenhout proclaimed that "the Linnean system of Nature is now too universally adopted to require any defense or apology," he added that "if it be not the most natural, it is doubtless the most convenient."[91] William Borlase took "pleasure in acknowledging my obligations to him," but he also suggested that the Linnaean system still contained "a few obscurities and perhaps improprieties . . . yet . . . to be retouched."[92] And the preface to an appreciative late-eighteenth-century account of Linnaeus's animal classification more darkly hinted that despite his "transcendent merits," he had attracted "the malevolent opposition . . . of numerous detractors."[93]

Linnaeus's early reception in Britain had in fact been mixed. When he visited England soon after he had begun to make his scientific reputation, several eminent naturalists, including Sir Hans Sloane, whose collection ultimately became the foundation of the British Museum, were inclined to snub him on the grounds that "he wished to overturn the old systems, only to exalt his own name."[94] Later, complaints emerged about the volatility of Linnaeus's system of classification, which changed in each edition of the *Systema Naturae;* according to one English distillation of Buffon's natural history, "by comparing the fourth edition of Linnaeus's Systema Naturae with the tenth, we find man is no longer classed with the bat, but with the scaly lizard."[95] Some critics queried the very principles upon which Linnaeus's classification was based. In 1759 a reviewer who found Linnaean taxonomy generally "arbitrary . . . chimerical . . . and . . . ill-grounded" particularly objected to the grouping of the dog with foxes and wolves and the horse with other hoofed animals; he suggested instead that the dog should follow the horse in natural history as it did in ordinary roads and farmyards.[96] And even in the nineteenth century, when, for most naturalists, Linnaeus had become a figure sufficiently remote to be revered and disregarded, he still occasionally aroused strong negative passions. For example, in his own work on mammalian classification, rather than piously claiming Linnaeus as an ancestor, William Swainson dismissed him as "radically wrong"; according to John Fleming, who associated the master with slavish disciples like Shaw, "the dogmas of the Linnean School" had been "conspicuously hurtful" and had "directly retarded the progress of Zoology in Britain."[97]

These expressions of antagonism derived from various sources. Even during the heyday of Enlightenment classification, gaps in the epistemic zeitgeist apparently left room for a lot of free-floating resistance to the very idea of system. Much of this resistance, especially on the part of naturalists and others uneasy with the intellectual distance that systematic classification interposed between the observer and the creature observed, crystallized around the renowned French naturalist Buffon. His voluminous and appealingly readable natural history was much more widely available in English translations and adaptations than was the uncompromisingly technical work of Linnaeus. Buffon was well known among naturalists as "the greatest enemy to Arrangement" in general and a severe critic of Linnaeus in particular.[98] In Buffon's view, as mediated to the anglophone reading public, "Nature . . . offers herself . . . in contradiction to our denominations and characters, and amazes more by her exceptions than by her laws."[99] His translator Oliver Goldsmith similarly declared that "saying an animal is of this or that kind is but a very trifling part of its history." He disparaged, for example, the systematic grouping of the hare and the porcupine "merely . . . from a similitude in the fore-teeth" on the grounds that this "slight" resemblance obviated much more significant differences, which he identified as "no likeness in the internal conformation; no similitude in nature, in habitudes, or disposition."[100] Later revisionists made similar arguments while dispensing with Buffonian support. Thus an early Victorian account of primates claimed that "minds . . . not too strongly enveloped in the trammels of system" could readily perceive that their manual dexterity and mental power entitled opossums to be classed with monkeys and lemurs.[101]

Resistance to the juggernaut of classification might also be obliquely expressed by means of an alternative format, the alphabetically organized dictionary or encyclopedia. Although dictionarists and systematizers shared an obvious goal—the pigeonholing of innumerable bits of information so that they would be easily retrievable—the logic of alphabetization, which tacitly recalled the outmoded bestiaries, ran counter to the logic of even the most arbitrary and artificial system devised by naturalists.[102] In his *New Dictionary of Natural History* William Frederic Martyn proclaimed that given "the sublime disorder of Nature herself, too prolific to enumerate or arrange . . . , and the essential variations between the most celebrated Naturalists, who confound while they attempt to explain . . . in the present work, we have emancipated ourselves from system."[103] Perhaps as a culminating gesture of defiance, his volumes lacked pagination as well as taxonomic structure. The anonymous author of a *Dictionary of Natural History,* published in 1802, echoed both Martyn's sentiments and his practice: "In most of the arts and sciences . . . there is a necessity for a systematic arrangement . . . But in natural history, this is by no means the case . . . Concerning the futility of systems in this part of literature, every person conversant herewith must be satisfied."[104] Even a devout naturalist like John Bigland, who felt that "some knowledge of the system of nature is necessary to *all ranks* of people," to help them advance both their material interests and their spiritual understanding, could sympathize with this reluctance to burden at least the nonspecialist audience with the machinery of classification: "The best mode of communicating useful instruction is to render it entertaining . . . Systematic arrangements, however advantageous . . . to the professional naturalist, tend more frequently to embarrass than to inform the juvenile student or the common reader."[105] And a popular guide to the newly opened menagerie of the Zoological Society of London implicitly made a similar point several decades later, noting that its "subjects, for obvious reasons, do not occur in the order of their scientific arrangement; but, by reference to the Alphabetical Contents [so that] the reader will be enabled to turn at once to the history of any animal."[106]

Such resistance to order, while vigorous, was not always absolute. In one way or another, alphabetical catalogues of animals usually acknowledged the widespread acceptance of systematic classification by incorporating it, albeit in a subordinate and therefore disparaged role. Even Martyn's rejection of system was less radical than his opening declaration and his alphabetical arrangement suggested. The entry for "Animal" included appreciative summaries of the work of such systematists as Ray, Klein, Brisson, and Linnaeus; that for "Quadrupeds" referred approvingly to Pennant's division of them into "digitated, hoofed, pinnated, and winged." He defined many terms drawn from and constituted by technical zoological nomenclature— for example, "Agriae" ("an order of quadrupeds destitute of teeth, but furnished with long cylindrical tongues") and "Glis" ("in a limited sense . . . only . . . the dormouse; but according to Linnaeus, the Glires constitute the fourth order of the mammalia").[107] A thoroughly consistent antisystematist would have found the more abstract or general nouns in these quotations not only unworthy of definition but devoid of meaning.

This grumbling uneasiness could also reflect tensions more loosely relevant to zoological debate. The inescapable analogy between the intellectual comprehension of nature and the practical management of people and territory encouraged some naturalists to interpret Linnaean classification as a species of intrusive alien authority—relatively easy to bear, as foreign yokes went, but a foreign yoke nonetheless. Criticism of Linnaeus leveled from this perspective transformed him from the preeminent representative of the supranational community of naturalists to a usurping carpetbagger. The resentment shared by such critics implied no concomitant theoretical unity; there was no anti-Linnaean consensus on a desirable replacement. But whatever system any particular naturalist preferred to that of Linnaeus, it was likely to have a British originator or forebear, rather than one from across the Channel.

The favorite candidate of patriotic naturalists was John Ray, frequently referred to as "our countryman" or "our illustrious countryman," and occasionally as "the Father of Natural History" or "the Aristotle of England." The stalwart John Fleming insisted that Ray was the father "not only of British, but of European natural history," and, in addition, that Linnaeus's British disciples had "suffered for their folly, in preferring the naturalist of Sweden" to one of their own fellow citizens, "vastly his superior in philosophical attainment, enlarged views, and . . . good taste."[108] It was often suggested that Linnaeus had somehow usurped credit due to Ray; thus a student in John Walker's natural history class at the University of Edinburgh dutifully recorded that "Mr. Rae was the first systematic [and] . . . Linnaeus followed [him] most."[109] Many zoological authors opted to follow Ray's lead rather than that of Linnaeus in arranging their works—although they offered no consistent explanation for their decisions. Richard Brookes claimed that "no systematical writer has been more happy [than Ray] . . . in reducing natural history into a form, at once the shortest yet most comprehensive."[110] Gilbert White, the clerical author of the *Natural History of Selborne*, felt that "foreign systematics are . . . much too vague . . . but our countryman, the excellent Mr. Ray is the only describer that conveys some precise idea in every term."[111] At the beginning of the nineteenth century, William Wood based his choice on moral grounds, noting defiantly that "rigid naturalists" might not be pleased with his decision, but hoping that "we may perhaps be pardoned for the repugnance we feel to place the monkey at the head of the brute creation [as Linnaeus had done], and thus to associate him . . . with man."[112]

Not every naturalist looking for local roots picked Ray. Some preferred the disciple to the master; for example, John Berkenhout asserted that "Mr. Pennant's System I think preferable to that of Linnaeus."[113] The more anatomically inclined might trace their classificatory lineage back to the distinguished eighteenth-century surgeon John Hunter, who had amassed an impressive museum of comparative anatomy in which the specimens were grouped according to organ systems. Some naturalists could find no favorite compatriot, and so had to be content with lamenting the absence; as one popular survey of natural history put it at the beginning of the nineteenth century, "to this moment, the British nation has not produced a systematic arrangement of living nature."[114]

Natural Systems

Nationalistic commitments added further complexity to a technical debate that was already vexed, slippery, and divisive.[115] Whether or not they appreciated Linnaeus's work, it became increasingly clear to eighteenth-century British naturalists—as, indeed, it had been clear to Linnaeus himself—that his system was artificial, in the sense that it tended to group animals on the basis of single characteristics, such as "dentition or the form of feet," often selected largely for classificatory convenience.[116] The alternative, referred to as a "natural system" and associated in the first place with John Ray, would ideally take into account a range of information about each organism and thus generate systematic categories that reflected the subtle and complex order of nature itself.[117] By the early nineteenth century the superior desirability of a natural system had become widely accepted, at least in principle, although some literalistic devotees of Linnaeus remained in influential positions, including the draconian George Shaw at the British Museum and the guiding spirits of the Linnean Society, who barred all discussion of these (or indeed any) issues at their meetings for fear of unseemly controversy.[118]

But it turned out to be much easier to acknowledge the need for a natural system than actually to devise one; as J. E. Bicheno commented in 1827, during the period when this issue was under the most intense and antagonistic discussion, "the difficulties of the subject have not been duly appreciated."[119] Prefatory to propounding his own rather unconventional systematic ideas, William MacLeay similarly noted that "if . . . there are few persons who form an accurate idea of what is meant by a System in natural history, . . . there are still fewer persons who comprehend the exact difference between the Natural and the Artificial System, although the whole science now depends on this distinction being thoroughly understood."[120] On a less theoretical level, zoologists attempting to elaborate a natural classification immediately encountered the problem that artificial systems so neatly evaded. Each animal had too many characteristics for them all to be weighted equally; failure to discern or establish the "natural" hierarchy of attributes would produce a more complicated and less straightforward system than that associated with Linnaeus, but one that in the end was no less artificial. Hotly contested though this debate was, therefore, some naturalists were inclined to abandon it altogether. As James Rennie argued in a handbook intended for zoological novices, albeit rather sophisticated novices, "every system must be artificial, inasmuch as it is only an effort of logical art in arranging in the mind a number of things which have many circumstances of resemblance and difference; and therefore all that has been said of . . . *a* or *the Natural System* is only a waste of words and an exhibition of critical fancies."[121] Darwin similarly suggested to a correspondent that "our ignorance of what we are searching after in our natural classifications" made it quixotic to view classification as more than "a simple logical process."[122]

Many taxonomists found it especially difficult to accommodate particular resemblances that apparently connected animals that otherwise seemed quite distinct. For example, Fleming somewhat oddly claimed that although the hare and the rabbit agreed "very closely in the structure and functions of their organs of protection, sensation, locomotion, and nutrition, and in these respects suitably belong to the same natural family," they differed so widely in the gestation of their offspring that "the hare . . . exhibits an affinity with the horse or sheep; while the rabbit . . . claims kindred (as does also the cat) with the opossum and kangaroo."[123] Schemata designed to represent all these perceived or potential similarities were apt to be both complicated and eccentric; and since their elusive object remained stubbornly unrealized, there was always room for another attempt. Thus, even though, in the course of the nineteenth century, the anatomical methods of classification associated with Georges Cuvier and Richard Owen became increasingly authoritative, there was a protracted period during which individual naturalists felt free to exercise a great deal of systematic discretion. And this privilege of choice was not restricted to the zoological elite or to the heights of theory; even the humblest laborers in taxonomic vineyards could consult their own judgment on practical applications. Writing in 1848 of Manchester weavers who devoted their leisure time to natural history, the novelist Elizabeth Gaskell noted that "there are botanists among them, equally familiar with the Linnaean or the Natural system."[124] For reasons that combined the scientific and the social, Linnaeus remained a live option for working-class naturalists long after their establishment colleagues had respectfully shelved him.[125]

Worth a Thousand Words

As these systematic alternatives proliferated, so too did the graphic metaphors or abstractions that at once represented them more clearly and persuasively than did mere language and ineluctably exposed inconsistencies that might otherwise have been cloaked by words. Throughout the eighteenth century, the dominant visual metaphor of system was the chain of being, or the *scala naturae,* an ancient figure that organized nature as a linked, one-dimensional progression from the meanest animal (or vegetable or mineral, depending on the perspective of the systematist) all the way to humans (or even to heavenly spirits, also depending on the perspective of the systematist).[126] By the end of the century the chain of being had become so ingrained in zoological discussion that it could be used axiomatically, as the basis for further theorizing or interpretation. Thus in 1802 the veterinarian Delabère Blaine explained his observation that "the higher order of animals, as more perfect machines, are less tenacious of life," with reference to the "beautiful gradating whole, blended with the nicest harmony into a vast chain" that constituted "animated nature."[127]

The imagery of the chain had become a ubiquitous feature of the language of zoological classification, especially in characterizing the relation between species or groups of species: thus the "Mouflon . . . may be considered as . . . forming the link" between sheep and goats, as, on increasingly abstract taxonomic levels, "the Hunting Leopard [cheetah] forms a sort of connecting link" between cats and dogs, the swine genus "may . . . form . . .

a link between the cloven-footed, the whole-footed, and the digitated quadrupeds," and the scaly manises [pangolins] "approach so nearly the genus
of Lizards, as to be the links in the chain of beings which connect the proper
quadrupeds with the reptile class."[128]

Although the authority of the chain was thus constantly reinforced by
explicit as well as by implied or figurative reiteration, its conceptual limitations became increasingly clear and burdensome to naturalists during this
period. The framework it provided was literally too strait and narrow to
accommodate the burgeoning number of animal species and the increasingly complex relationships perceived to exist among them. A carelessly
constructed chain could feature some alarming linkages, as Thomas Beddoes
demonstrated in a paper read to the Edinburgh Society for Investigating
Natural History in 1785. He pointed out that if a naturalist were to concentrate on nails and external covering (scales and shells) "a chain might be
constructed . . . the Links: Homo sapiens, Bradypus [sloth], Myrmeco-
phagus [anteater], Manis, Dasypus [armadillo], Testudo [tortoise] . . .
Whence it appears that we might get from an Alderman to a Turtle at 5
steps."[129] Even in the usage of loyalists, the metaphorical overtones of the
chain could easily slip from connection to confinement. In arranging his
Natural History of Cornwall, William Borlase warned that although "order,
connexion, rank, and relation must be strictly observed," nevertheless "there
must be no shackles."[130]

Or the chain might be complicated in ways that undermined or nullified
it. For example, Charles White conventionally asserted that "there is a
general gradation from man through the animal race," but he added that
"the gradation from man to animals is not by one way; the person and
actions descend to the orang-outang, but the voice to the birds."[131] The
chain was made flesh in the organization of John Hunter's comparative
anatomy collection, but in a radically disembodied or disemboweled form:
individual organs, rather than entire beings, were displayed in parallel linear
sequences—independent hierarchies, for example, of stomachs, genitalia,
and lungs.[132] As the outworn metaphor continued to chafe, more committed
antagonists made stronger criticisms. In the 1833 volume of the *Field Naturalist,* "Solitarius" characterized the chain as "nothing better than a mere
fancy"; a few years later Darwin confided to his notebook that "it is capable
of demonstration that all animals have never at any one time formed a
chain."[133]

Many naturalists dissatisfied with the chain offered alternative images.
After years of systematizing Linnaeus himself likened the living world to a

map, although his suggestion in this regard was not much attended to; Oliver Goldsmith compared the class of quadrupeds to a complex polygon, "terminated on every side by some that but in part deserve the name."[134] In 1841, giving voice to the emergence of a fresh scientific consensus—one that was part of a larger paradigm shift—Hugh Strickland asserted that "no *linear* arrangement . . . *can* express the true succession of affinities" and that while "symmetrical systems" were "agreeable to our love of order . . . we must not sacrifice truth to convenience." Using the newly dominant arboreal metaphor of system, Strickland urged his fellow naturalists "to study Nature simply as she exists—to follow her through the wild luxuriance of her ramifications, instead of pruning and distorting the tree of organic affinities into the formal symmetry of a clipped yew-tree."[135] Soon the tree became as conventional, in both visual and verbal form, as the chain had been. Within a generation, pedestrian elaborations like the following, taken from a review essay intended for a serious nonspecialist audience, were common-places of systematic discussion: "animal life [may] be likened to a great tree with countless branches spreading widely from a common trunk, and drawing their origin from a common root, branches bearing all manner of flowers, every fashion of leaves, and all kinds of fruit."[136]

Interestingly, however, this metaphor was broached in the mode of argu-mentation rather than of exposition—to rebut, however blandly, the notion that "the animal creation could be a linear series." Such rebuttals were more frequent and persistent than might have been predicted from the Victorian ascendancy of the tree metaphor. For example, in 1878, Alexander MacAlister felt obliged to preface his vertebrate zoology textbook with the warning that that "it is not easy to give in linear series an adequate idea of the relation-ships" among animals; he followed this warning with a diagram that was canonically arboreal in form, if rather eccentric in content—grouping, for example, cattle with monkeys on one large branch, and rats with horses on another.[137] In 1884 an expert on taxidermy regretted that "purposes of convenience and reference" required the "linear arrangement" of stuffed museum specimens, although "it will not be necessary to point out that no *actual* linear arrangement can exist in nature."[138]

As it turned out, however, reports of the death, or even the displacement, of the chain were greatly exaggerated. Although fewer and fewer naturalists explicitly endorsed it as a systematic model, it continued to shape the language of almost everyone who discussed relationships among animal groups, especially since the notion of linkage also reflected late Victorian evolutionism. If, from one perspective, a new orthodoxy had been estab-lished, from another, that of usage or practice, this changing of the metaphorical guard was far less obvious. Thus, in 1853, an authority on ovine husbandry referred to "those miscellaneous kinds [of sheep], which, either from their prevalence or their peculiarities, fill up the links in the chain between the Argali and the Leicester"; in 1872, an authority on zoonoses (diseases transmitted to human beings from other animals) noted that, as far as "structure and functions" were concerned, humankind "constitute but an upper link in the zoological chain."[139] At about the same time, the anatomist and racial scientist Robert Knox explained human distinctiveness in terms of the ostensibly discarded metaphor: "The human family stands profoundly apart from all others, implying that in the great chain of being constituting nature's plan, some natural family filling up the link has disappeared."[140] Writing near the end of the nineteenth century, the sophisticated William Henry Flower could mix the rival metaphors without any apparent awareness of inconsistency: "it is proposed . . . to treat of the horse . . . as one link in a great chain, one term in a vast series, one twig of a mighty tree."[141]

Perhaps the most striking example of the multivocality of systematic discourse—of the scope that existed for naturalists to exercise their imagination and intellectual willfulness—was the debate about the quinary system, which flared and sputtered through the second quarter of the nineteenth century.[142] The brainchild of William MacLeay, who proposed it in his *Horae Entomologicae* of 1819, the quinary system was an elaborate and eccentric attempt to represent the complex, overlapping sets of resemblances among animals. From the quinary perspective, the compounded linearity of the taxonomic tree was as unsatisfactory as the simple linearity of the taxonomic chain, because it similarly constrained the number of formal connections between animals.[143] That is, it privileged similarities of what was known as "affinity"—anatomical likenesses—over similarities of what was known as "analogy"—primarily likenesses reflecting shared habits, such as the convergent aquatic adaptations of whales and fish. As MacLeay put it, his "first endeavour, after discovering the principal affinities, was to ascertain the connexion that might exist between the general structure of the animal and its manner of living."[144]

Quinary nature was much more intricately organized than nature in the shape of a tree; it was also, in its fussy way, neater and more predictable. Quinarians arranged animals within a set of embedded circles, each of which consisted of five subsidiary circles, each of which was in turn subdivided into five smaller ones, and so on.[145] Thus, MacLeay divided the circle of the class Mammalia into five orders: primates, ferae (carnivores), glires (rodents), ungulata (hoofed animals), and cetacea (marine mammals)—

along with, awkwardly, a sixth "transultant" order containing the ever-troublesome marsupials. An additional principle of organization helped to structure each group of five: one of its members was identified as "typical," one as "sub-typical" (these first two united under the rubric of "normal"), and the remaining three as "aberrant." In a manuscript that detailed a scheme of mammalian classification, MacLeay identified the cats as the typical ferae, the hyenas as subtypical, and the weasels, dogs, and civets as aberrant (and therefore, among other things, less carnivorous); he similarly subdivided the bimana (in his view, although not in the view of every quinarian, the human group within primates) into the typical Europeans, the subtypical Asiatics (both "normal" because "civilized"), and the aberrant or "savage" Americans, Africans, and Malays.

Several other kinds of connections between remoter animal groups were most effectively indicated by graphic representation or mapping. Every circle was drawn touching other circles of the same taxonomic rank at each of its five significant points. These abutments indicated analogic bonds, so that, according to MacLeay, mammals were connected to birds through the conjunction of the glires with gallinaceous fowl (chickens and their kin) and to fish through the conjunction of the cetacea with sharks. Finally, each circle repeated the same abstract pattern, so that theoretically any two circles, whether at the same taxonomic level or not, could be superimposed to yield a set of equivalences. Thus MacLeay related the primates to the ferae as follows: humans were analogous to the plantigrade carnivora (those that walked on the soles of their feet), apes to the digitigrade carnivora (those that walked on their toes), lemurs to bats, and tarsiers to insectivores. He noted that "these analogies explain how some years ago, a Bear was exhibited to the London public as a wild Indian."[146] There was also an empty or null category in primates corresponding to the amphibious carnivora (seals). The postulated existence of such place markers thus gave the system some predictive as well as analytic force, suggesting that those categories would eventually be filled by creatures that must exist but had yet to be discovered.

At first, discussion of MacLeay's ideas was primarily limited to a small circle of sympathetic spirits who belonged to the Zoological Club of the Linnean Society. Many of the papers presented at club meetings dealt with quinarian topics, and the minutes often recorded lengthy corroborating testimonials from the floor.[147] But if the quinary system had remained within these sheltered confines it would have figured only as an index of the intellectual idiosyncrasies of MacLeay and his friends. The Zoological Club was hardly an obvious launching pad for influential new theories, and indeed, in the autumn of 1829, its dwindling membership voted to disband because fewer and fewer of them were bothering to turn up for meetings.[148] Gradually, however, quinarianism attracted wider attention; its dissemination and influence peaked in the 1830s, after it was taken up by William Swainson, a preternaturally productive author of popular works about natural history.[149]

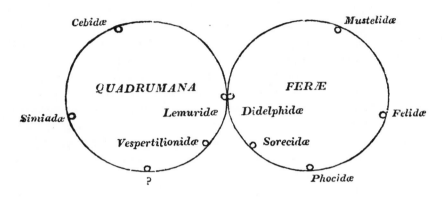

Some of Swainson's quinary circles.

Not surprisingly, much of this attention was negative. Diverse naturalists who nevertheless shared an appreciation of what John Fleming termed "the high expectations concerning the future progress of the science [natural history] in this country" tended to disparage MacLeay's system as a crackpot exercise in mythmaking. Although Fleming referred respectfully to MacLeay as "this intelligent naturalist," he suggested that his "novel method . . . seems to have originated in metaphysical prejudices, and by overlooking the fact that, in the various organs . . . belonging to each species, there are characters which enable the physiologist to trace resemblances in structure and function with the organs of many other species."[150] Thomas Rhymer Jones, who professed comparative anatomy at the University of London and physiology at the Royal Institution, criticized the quinarians in similarly condescending terms: "needlessly to . . . arrange trivial groups of external forms in imaginary circles, is an easy occupation to the superficial zoologist,—easier perhaps than it would be to one more deeply conversant with . . . anatomy."[151] Darwin, perhaps recognizing the threat it posed to his embryonic evolutionary theory, denounced quinarian systematics as "vicious circles" and "rigmaroles" in his correspondence and notebooks.[152] And after a few decades of skirmishing, the battle had been won, or so it seemed to these critics; MacLeay's system had been safely pushed beyond the pale of scientific respectability. By 1847, H. N. Turner could consign what he indulgently termed "this enchanting dream" to the irrecoverable past—"a time not long gone by."[153]

It may have been simple magnanimity to a fallen foe that led Turner to characterize MacLeay's adherents as including "some of the most eminent [naturalists] of their day."[154] But there were other indications that quinarianism was a more substantial scientific presence than the airy dismissiveness of some of its triumphalist rivals implied. It was significant that anti-quinarians felt a need to repeat their dismissals rather frequently over a period of several decades; the relatively impassioned language of some of the critics, such as Darwin, also suggested a seriously troublesome target. Most telling, however, was that fact that at the same time that some establishment naturalists were disparaging quinarianism, others were describing it in terms that were respectful and even admiring. According to the *Zoological Magazine*, it "may be contrasted, without the least disadvantage, with the classifications of the best modern naturalists"; and J. E. Gray, the Keeper of Natural History at the British Museum, published a quinary classification of the mammalia, along the lines indicated in MacLeay's draft.[155] In his progress report on zoology, presented to the fourth annual meeting of the British Association, Leonard Jenyns devoted more space to MacLeay's taxonomic system than to that of Cuvier, particularly noting the "influence which this theory has had over our own naturalists." Although he remained skeptical, he acknowledged that MacLeay and his adherents had "certainly advanced our knowledge of natural groups" and concluded that "it is difficult to believe that there is not some truth at the bottom of this theory, however erroneous it may be in its details."[156] Richard Owen characterized the ideas "which have emanated from the naturalists of the English Quinary school" as "remarkable for their novelty and boldness."[157] The young Thomas Henry Huxley similarly found it suggestive, if not convincing.[158]

Both theoretical praise of the quinarian system and practical application of its principles found their way into many of the manuals and other sources of zoological information that targeted the general public. In addition to Swainson's steady quinarian stream, for example, a contemporary agricultural encyclopedia noted that "the brilliant anatomical and physiological discoveries of Cuvier, Lamark, Latreille, and others . . . have laid the foundation of this [new] system; but it was reserved for our own countryman, Macleay, to generalise their details . . . this gifted writer has proved the existence of five primary divisions in the animal kingdom"; a guide to British animals boasted that an "arrangement of the Mammalia has been drawn up from a combined view of the system of Cuvier, and the systems of Gray and MacLeay."[159] *Vestiges of the Natural History of Creation,* which appeared anonymously in 1844, became notorious for its evolutionary or transformist argument; less controversially, it devoted an entire appreciative chapter to MacLeay's system.[160]

Explicit endorsements and subtle echoes continued to reverberate through the literature. As late as 1891, Essex projectors of a local natural history museum quoted the former Professor of Natural Science at the University College of Wales to the effect that in provincial collections "each group should be represented by the most typical and the least typical example . . . ; by a specimen taken from the centre, and a specimen or two near the circumference of the group where it is coterminous with another, or even overlaps it. Thus the . . . carnivora might be represented, not only by a dog and a cat . . . as central types, but also by a seal, . . . taken from one of the margins of the group where it abuts upon the whales."[161] And even long after quinary ideas had become the primary property of zoological eccentrics and interlopers—people like John Ruskin, who, near the end of a life mostly devoted to cultural struggles, made the subversive suggestion that "we may be content with the pentagonal group of our dabchicks"—they retained the power to annoy and even infuriate canonical scientific authorities.[162] Thus in 1875 *Nature* editorially blistered an unspecified author for writing "unmistakable, undeniable nonsense" about "that silly 'circular system' with its mystical numbers, its fives or its sevens—the will-o'-the wisp of fancy that once led men's minds astray from the path where only they could find the truth they were earnestly seeking."[163]

The Part for the Whole

The quinary system may not have been the most radical challenge to the systematic enterprise to emerge from within the zoological community. After all, it was recognizably the same kind of system as those produced by professed followers of Ray or Cuvier: the categories were familiar, however odd their arrangement, and MacLeay and his cohort shared with their more conventional or mainstream contemporaries the desire to devise a natural, rather than an artificial, classification—one that would take into account as much of the information that they deemed significant as possible. Less frequently targeted for criticism was a series of attempts, not linked explicitly or at all to the "artificiality" of hyperorthodox Linnaeans, to base taxonomies not on the whole animal but on constituent organs or tissues. John Hunter, whose name and museum were most persistently associated with this practice, had offered proliferating alternatives. In the words of his son-in-law Everard Home, Hunter arranged his unrivaled anatomical collection according to "a system of his own, . . . which bears no resemblance to those of other naturalists." It was "an attempt to class animals according to their . . . internal organs" through series of increasingly complex hearts, stomachs, and other viscera; more specialized and idiosyncratic series, such as "organs of progressive motion adapted for flying" and "eyes modified for seeing in water," were also included.[164] The same creature could thus have very different neighbors in different parts of the museum: the kangaroo's hair classed it with the rabbit, its upper jaw with the hog, and its back toes with the ostrich.[165] This profusion of inconsistent affinity groups, however rigidly structured each was internally, inevitably suggested that organizing the natural world was a matter of choosing among attractive options rather than discerning a single inherent pattern.

Many like-minded colleagues less expansively preferred to identify a single attribute, function, or organ from which a complete taxonomy of mammals could be derived. But the number of such attempts and their lack of agreement about which characteristic was diagnostic meant that their aggregated classifications made the same unsettling point as Hunter's multiple one. At about the same period that Ray was beginning to explore the possibility of a natural system, Nehemiah Grew was convinced by his wide experience of dissection that the gut possessed primary taxonomic significance. He therefore suggested dividing quadrupeds according to what they ate, into the carnivorous (weasel, dog, cat), the insectivorous (mole), the frugivorous (hedgehog, mouse), the frugivorous and graminivorous (rabbit, pig, horse), and the graminivorous (sheep, calf).[166] Along similar lines, over a century later, John Walker, the professor of natural history at the University of Edinburgh, told his students that "the Mammalia may be distinguished by food," and the author of an introductory zoological guidebook revealed that "Quadrupeds . . . are divisible into *Carnivorous, Herbivorous,* and *Mixed* or general feeders."[167]

The digestive tract was not, however, the favorite focus of unidimensional taxonomies. Instead, naturalists searching for a classificatory touchstone tended to concentrate on functions that they associated more closely with an organism's essential nature. For example, some zoologists were inclined to differentiate on the basis of mentation and therefore focused on the shape of the head, so that one critic taxed all the eighteenth-century systematists for not having "ascertained the forms of heads [of different animals] by a regular standard fixed on mathematical principles"; a paper read to the Wernerian Natural History Society of Edinburgh in 1816 was entitled, "Observations on the possibility of distinguishing the characters of animals by the form of their crania, made with a view to establish a system of zoology founded on the forms of the brain."[168] Others less phrenologically oriented chose to probe the brain itself. The Linnean Society *Transactions* of 1837 contained a division of the placental mammals according to the number of their cerebral lobes into Educabilia (primates, carnivores, marine mammals, and hoofed animals), happily endowed with two or three, and the less fortunate Ineducabilia (edentates, bats, rodents, and insectivores), which had only one. Two decades later Richard Owen detailed his "fourfold primary division of the mammalian class, based upon the four leading modifications of cerebral structure" before the same institution.[169]

The low road of sex attracted as much attention as the high road of intellect, although choosing it was more problematic in terms of public relations. As one late-eighteenth-century naturalist lamented, "an attentive investigation of the genital organs might offend, though from the various structure of the clytoris, nymphae, scrotum, and penis, considerable service might be derived in forming a natural arrangement of animals."[170] Undismayed, however, subsequent zoologists boldly continued to suggest variations on this theme, most frequently using the ovum or the placenta as the basis for their classification.[171]

Most of those who proposed organ-based systems identified themselves primarily as comparative anatomists. Despite the increasing reliance of taxonomists on anatomical and physiological criteria, and the largely overlapping subject matter of anatomy and natural history, significant tension existed between these disciplines. Although, from a broad perspective, as Thomas Henry Huxley explained to a group of student teachers in 1860, their interconnection was clear—"classification is the expression of the relationship which different animals bear to one another, in respect of their anatomy and their development"—many workers in the trenches seemed rather to feel competitive and even hostile.[172] Thus in his characteristically exaggerated idiom John Ruskin, a very amateur naturalist and no friend of zoological taxonomy either, told his Oxford students that "anatomical study . . . has, to our much degradation and misfortune, usurped the place, and taken the name, at once of art and of natural history."[173]

Anatomists tended to express their reciprocal convictions more moderately, implying that their advanced analytical purposes required a different sort of system than satisfied the needs of mere descriptive zoology. Earlier in the nineteenth century, at the beginning of a course on comparative anatomy, William Lawrence offered a classic formulation of this position, announcing that the organization of his lectures would be "physiological;

... depending on the arrangement ... of the organs or functions. In natural history . . . the species are described separately: this plan is altogether unsuited to our present purpose, as it would be attended with endless repetitions, and would exhibit the facts in so detached . . . a form, that very little use could be made of them."[174] Or, as Home said of Hunter's collection, it was "little favourable to the study of natural history, but is peculiarly adapted for . . . comparative anatomy."[175] In this way a radically alternative principle of classification could become the banner of internecine disciplinary or subdisciplinary rivalry. Long after the heroic labors of the founding systematists, and even within the institution of science, a great deal remained up for grabs. To participants in a project conventionally hailed as demonstrating the intellectual conquest of nature, the internal history of zoological classification might seem as much a constantly shifting kaleidoscope of competing systems and principles as a steady evolution and elaboration of a dominant paradigm.

Different Strokes

Outside the community of naturalists, rival authorities were still more numerous. After all, learned specialists had never been the sole, or even the primary, source of expertise about animals. The wisdom accumulated by those who worked with them every day, whether farmers or hunters, trainers or butchers, had been valued by the authors of Renaissance compendia. By the end of the seventeenth century, however, naturalists eager to emphasize the relative sophistication of their own expertise increasingly disparaged what they regarded as the "folk tales" of an ignorant peasantry,[176] much as their successors would defensively disparage the bestiary tradition. As in the bestiary case, however, the separation proved easier to assert than to demonstrate. When they were worked out in detail, even the taxonomic systems that embodied this assertion of learned preeminence did not unambiguously confirm that they represented a new departure in the study of animals, or that the theoretically informed discriminations of specialists were superior to—or even different from—those with explicitly pragmatic or traditional foundations.[177]

Often, on the contrary, eighteenth-century systems reflected competing, if unacknowledged, principles of organization that undermined both their schematic novelty and their claim to be based on objective analysis of the natural world. These competing principles usually divided animals into groups based not on their physical characteristics but on subjective perceptions of them. An ancient category like vermin—animals harmful to human interests and therefore "necessary to be hunted," as one seventeenth-century manual of venery put it—survived in the taxonomy of Ray and Pennant, only superficially transfigured by the claim that it alluded to the sinuous forms of the weasels, martens, and polecats most frequently so castigated.[178] Rather than analyzing nature exclusively on its own terms—the claim embodied in their formal systems—naturalists often implicitly presented it in terms of its relationship to people, even constructing formal categories that echoed the anthropocentric and sentimental projection characteristic of both the bestiary tradition they had so emphatically discarded and (then as now) of much vernacular discourse about animals.[179]

Richard Brookes introduced one of the most powerful and highly constructed sets of subjective categories in *The Natural History of Quadrupeds*. He proclaimed himself a follower of Ray, and therefore, like most naturalists in this camp, he used a range of physical characteristics to group mammals, with foot conformation preeminent.[180] This method yielded the following arrangement: horses (undivided hoofs); ruminants (cloven hoofs); the hippopotamus, elephant, and others (anomalous hoofs); camels; monkeys (the first of the animals without hoofs); humans; cats; dogs; weasels; hares; the hedgehog, armadillo, and mole (divided feet and long snouts); bats; and, finally, sloths. Yet he also asserted, and this in the synoptic introduction where his methodological consciousness might be presumed to be highest, that "the most obvious and simple division . . . of Quadrupedes, is into the Domestic and Savage."[181] Obvious though it might have been, however, this division implied a taxonomic structure that cut across the formal organization established in Brookes's synoptic summary (unless, perhaps, that summary was taken to represent a progression from familiar to exotic animals, rather than an analysis of mammals based on feet). A more forthright contemporary, who disliked most Enlightenment systems, explicitly proposed this dichotomy as an alternative source of the "method, arrangement, or classing" that he nevertheless acknowledged to be required by natural history.[182]

The distinction between animals who served people and those who remained beyond human influence was not inevitably subversive of zoological systematizing. The differences, for example, between wild animals and the domesticated creatures that most closely resembled them could fit easily into routine subgeneric analysis. The uncompromisingly Linnaean George Shaw defined the "common ox" as "the Bison reduced to a domestic state, in which . . . it runs into . . . many varieties . . . differing widely in size, form, and colour."[183] Observed over the range of domestic species, however, such characteristics invited generalization, and these generalizations might lead to the formation of categories more difficult to assimilate into a system based on the principles of Linnaeus, Ray, or even Buffon. It was commonplace to notice that, as one retailer of Buffon put it, "domestic animals in very few respects resemble wild ones; their nature, their size, and their form are less constant . . . especially in the exterior parts of the body."[184] Related, if less firmly anchored in objective observation, was the assertion that "all Animals, except ourselves . . . are strangers to pain and sickness . . . We speak of wild animals only. Those that are tame . . . partake of our miseries."[185] A Victorian agricultural authority transformed differential morbidity into a critique of domestication itself, claiming it had "contributed to [the] degeneracy" of animals, which, in their wild state, "were at first perfect."[186]

And often, like Brookes, zoological writers treated the distinction between wild and domestic animals as primary rather than contingent, the basis for taxonomic discrimination rather than the occasion for description and explanation. Thus William Swainson defined domestication not as a human accomplishment but as an innate, divinely inculcated propensity "to submit . . . cheerfully and willingly." In consequence, it had systematic significance, cropping up in "every instance among the more perfect animals [of] the rasorial type."[187] Many naturalists used domestic animals as taxonomic models, typically claiming that "each class of quadrupeds may be ranged under some one of the domestic kinds"; thus domestic animals both exemplified and limited the range of mammalian possibilities.[188] Oliver Goldsmith, among others, made this assertion even though the resulting system left many animals uncategorized—elephants, hippopotami, giraffes, camels, bears, tapirs, badgers, and sloths, to name a few. The categories of wild and domestic might be perceived as so disparate that they required a connecting link: for example, "as the cat may be said to be only half domestic; he forms the shade between the real wild and real domestic animals."[189]

The dichotomy between wild and domestic animals was equally influential at lower taxonomic levels. In the Naturalist's Library account of ruminants, William Jardine provided separate entries on the reindeer of North America and those of Eurasia, not because he doubted that they belonged to a single species but because "in the one he is hunted in a state of nature, while in the other the greater proportion of the race is under the guidance and protection of man."[190] Similarly privileging relationship to humanity over physical characteristics, he revised a colleague's placement of the "subgenus *Taurus* [that containing domestic cattle] last in the series of Bovine Animals. We have treated it first, as containing animals of the most importance."[191]

Another indication of the taxonomic significance of domestication was that animals perceived to differ only in this attribute—that is, wild animals that were anatomically similar to domesticated ones and that were believed to interbreed with them—were frequently placed in separate species. Pennant, along with many others, identified the European wild cat as the "stock and origin" of the domestic cat, but one was *Felis sylvestris* and the other was *Felis catus.*[192] Indeed, the fact of domestication appeared so important to some naturalists that they tended to accord breeds of domestic animals the same taxonomical status as species of wild animals, despite the fact that variability within species was widely recognized as one of the most frequent consequences of domestication. In his *General History of Quadrupeds,* Bewick presented "The Arabian Horse," "The Race-Horse," "The Hunter," "The Black Horse," and "The Common Cart Horse" in separate entries analogous to those devoted to "The Ass" and "The Zebra"; in his *Cabinet of Quadrupeds,* John Church equivalently offered illustrations of both the "Barbary Ape and Orang-Outang" and the "Bull Dog and Pomeranian Dog."[193] Following Linnaeus, Shaw tagged many dog breeds with latinate binomials that at least sounded like the names of species—for example, the hound was *Canis sagax,* the shepherd's dog was *Canis domesticus,* and the pomeranian was *Canis pomeranus.*[194] And if divisions based on the dichotomy between wild and domestic animals came into explicit conflict with divisions based on the principles of systematic taxonomy, sometimes it was the latter that gave way. When Edward Bennett admitted that "it would . . . appear . . . impossible to offer" a physical description of the domestic dog that would distinguish it from the wolf and other wild canines, he did not conclude that they should all be considered a single species. Instead, to reify the division based on domestication, he introduced a new and circular criterion: "it is to the moral and intellectual qualities of the dog that we must look for those remarkable peculiarities which distinguish him."[195]

Nor were wildness and domestication the only taxonomic differentiae imported—or smuggled—into systematic zoology from a lay world of utilitarian and anthropocentric discriminations. Some were anthropomorphic as well as anthropocentric; systems were devised that not only classified animals on the basis, one way or another, of their relationship to people, but that classified them in terms that, unlike wildness and domestication, had developed primarily or exclusively to distinguish human groups. Eighteenth- and nineteenth-century zoological catalogues repeatedly rehearsed the categories of political geography.[196] The artist Thomas Rowlandson labeled a series of engravings published in 1787—*Foreign and Domestic Animals*—with a rather reductive version of these categories, but most zoological authors were more precise.[197] The heartfelt wish expressed by Edward Tyson at the end of the seventeenth century for "a good History of the Animals of our own Countrey" was ultimately granted in spades with, among others, William Bingley's *Memoirs of British Quadrupeds* (1809), Leonard Jenyns's *Manual of British Vertebrate Animals* (1835), Thomas Bell's *History of British Quadrupeds* (1839 and 1874), and William MacGillivray's *British Quadrupeds* (1838), a part of the Naturalist's Library, which otherwise organized its volumes according to more orthodox taxonomic units.[198]

Nevertheless, at least in some quarters, the demand for such rehearsals of the national fauna continued unsated. The prospectus of the Ray Society, founded in 1844 to encourage high-end natural history publication, listed among its special desiderata "a systematic history of the zoology and botany of the British Islands," although Hugh Strickland, one of its early subscribers, demurred that "there is rather too much stress laid . . . upon *British* Natural History . . . of all branches of zoology the most *popular,* and therefore the best able to swim without corks."[199] And near the end of the century Richard Lydekker justified his publication of yet another *Hand-Book to the British Mammalia* on the grounds that no such work had appeared for twenty long years; at about the same time, and presumably in response to the same perceived demand, the editors of a magazine called the *Young Naturalist* rechristened it the *British Naturalist* and announced that it would "give most attention to the British Fauna."[200] Smaller political units were similarly reified as facts of nature. In his late Victorian *History of St. Ives,* Herbert Norris noted that "as there is no published list . . . of the Natural History of Huntingdonshire, I venture to give a list of those animals and birds I have personally observed"; a guidebook to York prepared for the seventy-fifth meeting of the British Association included a list of "the Mammals occurring within a twenty-mile radius of York"; each of the Victoria County histories contained a roster of the shire fauna.[201]

This ethnocentric taxonomy was also—and more concretely—enshrined in museum collections. In 1787 the Edinburgh Museum of Natural Curiosities advertised a collection "containing the whole of British Zoology" classed "in their proper order" as "a very great entertainment to the curious [and] an ornament to the city."[202] The London Museum and Institute of Natural History opened in 1807 as "a national academy of the natural history of our native land," its large collection "disposed in the order of Scientific Arrangement . . . and . . . calculated to display, in the most pleasing and impressive manner, the Grandeur, Variety, Beauty, and intrinsic Value of the native Riches of the Country."[203] In the museum maintained by the Zoological Society of London during its first years, according to the curator, "British species are distinguished from the Foreign by the English names being printed in red ink on the label."[204] Although the British Museum took the entire globe as its catchment area, it too gave special attention to British animals, which were sought as eagerly as those from exotic regions and much more comprehensively displayed.[205] In 1816, the keeper of its zoological department promised that the name of anyone who donated samples of "any species, sex, or accidental variety" absent from the collection would be "constantly registered." Among the creatures singled out as particularly desirable were the Shetland pony, the red ox, the wild goat, and the small-headed narwhal, all claimed as native because they had "undoubtedly been killed in Great Britain."[206]

Institutions of less than national significance might analogously restrict the *patriae* defined and illustrated by their natural history collections. A very good school museum could appropriately boast a nationally oriented assemblage of stuffed animals, but most schools, like most provincial museums, were advised to focus on "the natural history of their own neighbourhoods."[207] According to a local guidebook, in the Royal Albert Memorial Museum at Exeter "the local collections of Archaeology and Natural History are extensive and varied"; the Cardiff Naturalists' Society was founded specifically to provide local specimens for the Cardiff Museum.[208] At the Saffron Walden Museum in Essex, "the concentration of specimens peculiar to the District" was "a leading feature."[209]

That naturalists should find the zoological (and botanical and mineralogical) productions of their own neighborhood particularly interesting and attractive was unsurprising and, indeed, inevitable. Amateurs, especially those who had only modest means or intermittent attention to devote to their pastime, were apt to concentrate on what was closest to hand. As John Fleming put it in the introduction to his own comprehensive guidebook, "the study of British Zoology is peculiarly attractive . . . by the facility with which many species . . . can be procured."[210] Even among the magnificent and varied collections of the British Museum, the exhibit of British animals proved disproportionately appealing to the humbler visitors, piquing the curiosity of novices and supplementing the field observations of veteran naturalists, who often came to discover the scientific names of specimens they had captured.[211] Provincial museums and other improving institutions similarly capitalized on the affinity of their target audiences for the local fauna. They shared the strategy of a school zoology textbook that "aimed at conveying correct ideas of the peculiarities of structure by which the principal divisions of the animal kingdom are distinguished" by examining "the most common native species belonging to each group." Its "great object" was not merely to inculcate isolated facts but "to bring natural-history knowledge home to the personal experience of the pupil."[212]

This emphasis on the zoological body politic might seem to echo the emergence among elite or professional naturalists of the subdiscipline of biogeography, which was rooted in the surprised recognition by early European explorers that different parts of the world—most strikingly the Americas and Australia—had distinctive faunas and floras.[213] By the late nineteenth century, scientists had formally divided the globe into six more or less distinct regions, often denominated the Palaearctic (northern Eurasian), the Nearctic (North American), the Neotropical (South American), the Ethiopian, the Indian, and the Australian.[214] Charles Darwin devoted an entire chapter of *On the Origin of Species* to geographical distribution, with special attention to island fauna; for similar reasons Alfred Russel Wallace wrote a book about *The Study of Island Life*.[215] But biogeography turned out not to encourage special attention to the counties, nation, or even continent inhabited by Victorian naturalists. Great Britain was not the kind of isolated ocean island important to Darwin and Wallace. It was insufficiently separated from the European continent by either time or distance to merit much zoological interest; indeed, during every Pleistocene glaciation the English Channel had receded, leaving a land bridge in its stead. A late Victorian zoogeographical textbook deflatingly insisted that the distinction of native zoology lay "chiefly in the poverty of the British fauna as compared with that of the Continent."[216] And in any case, although biogeographers occasionally used the word *nation* to characterize regional assemblages of animals and plants, they ordinarily did so with explicit consciousness of the dangers of metaphorical slippage. Philip Lutley Sclater and his son William began their survey of mammalian biogeography with the warning that "it has long been evident to naturalists that the ordinary political divisions of the earth's surface do not correspond with those based on the geographical distribution of animal life," and then proceeded to give the most sobering possible example: "Europe . . . the most important of all the continents politically . . . is for zoological geographers . . . but a small fragment of Asia."[217]

Naturalists not specifically concerned with these issues tended, however, to disregard these warnings. They preferred to inscribe their favorite political categories in the book of nature. This predilection might lead them to identify a distinctively British fauna, complete with separate varieties and even species, which efforts in their turn inspired intermittent revisionist backlash, as when Leonard Jenyns read a paper to the Linnean Society entitled "Some Observations on the Common Bat . . . with an attempt to prove its Identity with the Pipistrelle of French Authors."[218] Or it might inspire a kind of zoological special pleading, as when a mid-nineteenth-century chronicler of Staffordshire natural history claimed that, although "the mammalia of Great Britain are limited in number [and] many . . . are very small," they were nevertheless "well worth . . . attention."[219] Some patriotic naturalists, following the example of Shakespeare's John of Gaunt, presented the sceptered isle as a "little world" or microcosm—"a most interesting epitome of the animals of this earth"—which could compete with the productions of the most remote and exotic regions, with no need for "foreign aid in acquiring the first principles . . . of zoological science."[220] One overview too modestly pointed out that, "with the exception of the apes and monkeys, there are few African quadrupeds that have not some relative, however humble, in these islands."[221]

Finally, the assumption that nature followed the flag frequently inclined naturalists to assess animals that they considered their fellow Britons as superior to alien beasts. This sense of added value conflated national feeling with other widely appreciated qualities. Some naturalists thus designated only wild species as British on grounds of pedigree or ancient descent, denigrating domestic animals as not truly native but "originally *introduced* into these Islands." Others based the same exclusion on grounds of character, as when the proprietor of a menagerie of British animals distinguished his show from "a mere collection of pets: for which *role* domestic animals . . . are more suitable."[222] The superior desirability of genuinely British animals could be more positively expressed in conventional cash terms, which tempted unscrupulous dealers retroactively to naturalize stuffed foreign specimens of animals that also occurred in Great Britain.[223]

The persistent willingness of zoologists to give systematic expression to such boundaries as those dividing the wild from the tame and the French from the British suggested that taxonomic advances did not trickle unimpeded down some clearly demarcated intellectual ladder. On the contrary, the delicate membrane that separated specialist and lay expertise was highly permeable in both directions. Further, it was not only with regard to relatively subtle or minor points of systematic analysis and interpretation that the authority of popular wisdom or pragmatic experience could prove an effective counterweight to that of zoology. The boundaries dividing the more inclusive or abstract taxonomic units, such as classes and orders, proved at least as controversial as those between similar species. And when contested issues of classification pitted evidence available for all to see against the more arcane analyses adduced by naturalists, no matter with how much consensus and force, many nonspecialists preferred to believe their own eyes.

Earth, Sea, and Sky

That Aristotle thousands of years previously had identified the taxonomic inadequacies of the biblical assignment of birds to the air, beasts to earth, and fish to the sea, if not precisely in those terms, was frequently pointed out by eighteenth- and nineteenth-century British naturalists who found themselves fighting the same battle. Resistance to the inclusion of flying and swimming creatures within the class Mammalia diminished only very gradually, even in the face of concerted and authoritative contradiction. The notion that the bat was a kind of bird should have been the easiest to dismiss, since, as Richard Brookes put it, "except the wings and flying, it has nothing in common with birds; whereas it agrees perfectly well in all things else with four footed animals."[224] Pennant succinctly summarized the basis of this confident classification: "no birds whatsoever are furnished with teeth, or bring forth their young alive, and suckle them."[225] Ringing though they were, however, such assertions failed in some cases to counteract the mute but powerful testimony provided by wings and buttressed by testimony from unreliable sources variously identified as *"Pliny, Gesner, Aldrovandus"* or, more vaguely and awkwardly, "some naturalists."[226] For if eighteenth- and nineteenth-century naturalists were unlikely to share the belief apparently still cherished by many of their readers that bats were birds, some were sufficiently impressed by the force of popular or traditional conviction to hedge their categorical bets. Church began his description of the bat by noting that it "seems to form a connecting link in the great chain of nature, between the Quadrupeds and the winged inhabitants of the air," even though, as was widely recognized by naturalists, "it is evidently related to the four-footed tribe, both by its internal and external formation"; Bewick agreed that the bat "seems to possess a middle nature between four-footed animals and birds"; and, in his somewhat more poetic vein, Chappelow characterized bats as "an uncouth, plumeless race of beastly birds."[227] A zoological survey published in 1844 by the Society for Promoting Christian Knowledge somewhat skeptically conceded that "they are classed among British quadrupeds; but are unlike any quadrupeds."[228]

Creatures of the deep seemed still more ambiguous than creatures of the air. Even among themselves, naturalists who discussed marine mammals were preoccupied with problems of classification. Cetaceans (whales and dolphins), pinnipeds (seals and walruses), and manatees were difficult to observe and to collect; in addition, their striking adaptations to aquatic life tended to overshadow the anatomical features that they shared with mainstream quadrupeds. Discerning their relationships to each other and to terrestrial mammals was consequently an extremely problematic and contentious task; for example, a *Zoological Magazine* author argued in 1833 that although the shape of the seal and its absence of legs appeared to ally it to the whale, "when its teeth and claws are attentively examined, these unerring guides distinctly point out its relation to the carnivorous order."[229] But when the universe of discourse was expanded to include those with less rarefied interest, much grosser taxonomic issues had to be addressed. If the "fish-like form" of the seal prompted some people to think the seal might be a whale, it prompted others to think it might be a fish.

Like those who reluctantly admitted the bat's relation to the "four-footed tribe," some naturalists were willing to gesture in the direction of taxonomic accommodation when speaking of marine mammals. Bewick asserted that although the walrus seemed "to partake greatly of the nature of fishes," it was "nevertheless classed by naturalists under the denomination of quadrupeds"; he even seemed to separate himself somewhat from this consensus

The talking and performing fish.

by calling it "the last step in the scale of Nature, by which we are conducted from one great division of the animal world to the other."[230] At the beginning of the nineteenth century, John Bigland similarly queried the judgment of his peers, noting that the walrus's "habits . . . approach nearer to fishes than to . . . quadrupeds, although naturalists have generally included it in the latter denomination."[231] The popular identification of pinnipeds as fish was still more durable. In 1854, the *Illustrated London News* published an article entitled "The Performing Fish," which described a remarkable animal (it could talk as well as obey orders) then on show under that rubric in London. The accompanying illustration, however, clearly displayed a seal, and, toward the end of the article, the author somewhat laxly addressed the taxonomic issue thus raised by suggesting that "it is . . . generally allowed that the term 'fish' is partly a misnomer."[232] As late as 1871 provincial museum-goers needed to be reminded that "seals . . . have no affinity with fishes," and a few years later, J. W. Clark began a public lecture at the London zoo with the bald statement that "the Pinnipedia . . . are true mammalian animals, entirely differing from fish both in structure and habit."[233]

Where cetaceans were concerned, this tendency to accommodate competing traditional categories was still more striking. It is true that zoological orthodoxy asserted that cetaceans were mammals: in Shaw's early-nineteenth-century formulation, "their whole internal structure resembles that of other Mammalia," or, as Tyson had more picturesquely put it a century earlier, "if we view a *Porpess* on the outside, there is nothing more . . . a fish; if we look within, there is nothing less."[234] The consensus inspired by these anatomically based assertions, however, seemed less than solid. Naturalists were almost eager to break ranks. Bewick probably assigned the role of mediating between mammals and fish to the walrus, because he considered whales and dolphins, which did not appear in his *General History of Quadrupeds,* to be incontestably fish. Pennant made the same omission in his *History of Quadrupeds,* and another serious late-eighteenth-century naturalist, John Reinhold Forster, explicitly classified the cetaceans as fish in his *Catalogue of the Animals of North America.*[235] In his rather personalized edition of Linnaeus, Robert Kerr complained that "this order of Cete ought, from external shape and habits of life, to have been arranged with the class of fishes, but the illustrious author having adopted the ingenious idea of employing the circumstance of shielding their young as a characteristic mark for a number of animals, all of which have warm blood . . . found himself forced to include these."[236] And Bigland, while acknowledging the justice of Linnaeus's categorization, converted this apparent strength into a critique, on the grounds that such counterintuitive taxonomy lacked mass appeal: "although he classes the whale species with the human and quadruped race, merely on account of the conformity of their teeth and the circumstance of being furnished with pectoral teats, this . . . will never prevent the whale from being considered as a fish, rather than as a beast, by the generality of mankind."[237]

Most naturalists hesitated to apply a democratic standard to taxonomic discriminations, and so, within the community of experts, sympathy for this heterodox view gradually evaporated. By the middle of the nineteenth century a combatant in other classificatory battles could confidently proclaim that "all systematists have agreed in placing these animals among Mammalia."[238] But this hard-won unity seemed to have little effect outside the charmed zoological circle; the "generality of mankind" continued stubbornly to adhere to its traditional commonsense opinion. In 1827 the young Charles Darwin was shocked to hear Henry Holland, a fashionable doctor and his distant relative, claim that whales were cold-blooded, and several years later an introductory zoology manual written by a professor at the University of London began its discussion of quadruped classification by excluding "lizards, frogs, whales, dolphins, and dugongs."[239] John Stuart Mill implied that limitations on the authority of scientific expertise were both inevitable and beneficial when he pointed out that although whales were mammals for zoological purposes, they were fish for commercial purposes; thus, in his view, "a plea that human laws which mention fish do not apply to whales, would be rejected at once by an intelligent judge."[240] At the end of the century zoologists who wished to discuss cetacean matters with nonspecialists routinely assumed that even audiences self-selected for their interest in such topics shared these outmoded convictions. Lecturing at the Royal Colonial Institute in 1895, William Flower rather truculently announced that "from the point of view of a zoologist . . . (whatever the fisherman and the man of business may continue to say) the whales and their allies belong to the class *Mammalia,* and not to that of *Pisces.*"[241] And in his *Royal Natural History* of the same year, Richard Lydekker stated with more resignation that "so like are Cetaceans in the general outward appearance to fishes, that they are commonly regarded as belonging to that class."[242]

That members of the late Victorian public stubbornly clung to their own opinions about the taxonomic status of whales and dolphins was just one indication of the multivocality of eighteenth- and nineteenth-century discourse about the natural world. During this period, naturalists added continually to the stock of information about animals; their constantly increasing authority was both distilled in and represented by successively refined systems of classification. The views of zoologists were ever more widely disseminated, both within the community of experts and to those with merely an amateur or even an idle interest in their subject. But whatever the hopes or intentions of the experts, this accumulation of theory and information was no hegemonic juggernaut, crushing error and dissension in its path. Even among naturalists there was room not only for scholarly debate but for deeply personal or eccentric alternative interpretations. And scientists were not the sole claimants of this zoological turf; British society included many rival sources of expertise, analogously symbolized by rival taxonomies. The persuasive force of alternative systems could be expressed passively, by the widespread reluctance to relinquish them, even in the face of overwhelming evidence, or more actively as overt rejection of authority. Perhaps its most striking manifestation, however, was the steady, if unacknowledged, incorporation of alternative points of view into canonical taxonomy. This series of interpenetrations inevitably undermined the binary opposition, earnestly posited by the emerging scientific establishment, between expert and lay knowledge. Instead of a dichotomy separating zoology from the rest of the culture to which its exponents belonged, the vigorous presence of competing ideas constructed a continuum of expertise about animals. Nowhere was this more apparent than in the language used to express and create systematic categories.

2

Flesh Made Word

B Y THE MIDDLE of the nineteenth century, most British naturalists
considered the classificatory principles of Linnaeus quaint and artificial. But
one major component of his work survived with its authority apparently
little diminished. As a committee of the British Association for the Advancement of Science reported in 1842 with regard to "the *binomial system of
nomenclature,* or that which indicates species by means of two Latin words,
the one generic, the other specific, . . . this invaluable method originated
solely with Linnaeus."[1] Like any system of classification, that of Linnaeus
required an associated terminology; and, as with any system, the relation
between categorization and naming was so close as to make it difficult to
distinguish between them. Yet, from the beginning, naturalists were inclined
to perceive Linnaean nomenclature as somewhat independent of the taxonomic scheme to which it ostensibly belonged. In making this distinction,
they may have been following the master's own lead. As early as 1736, in a
broadsheet inserted into the first edition of the *Systema Naturae,* Linnaeus
outlined the "method . . . by which the Physiologist can accurately and
successfully put together the history of each and every natural object" under
seven headings, of which the first and therefore most prominent was
"Names."[2] His achievements in the realm of nomenclature formed the basis
of his own most ambitious and memorable estimate of his work, his
self-designation as the second Adam.[3]

Once naturalists had separated Linnaeus's nomenclature from the rest of
his system, they tended to evaluate it much more favorably. In his *History
of Animals* (1752), John Hill attributed his rather precocious adoption of
Linnaean arrangement to the superior attractiveness of Linnaean nomen-

clature, as a result of which "those who study these subjects, had acquainted themselves with his new names, much better than the old."[4] Hill had no particular quarrel with Linnaean classification, but many of those with reservations about the system exempted the nomenclature from their general critique. Near the close of the eighteenth century, for example, Thomas Pennant emphatically rejected Linnaeus's arrangement of quadrupeds, primarily on grounds of inappropriate associations—he objected as much because "the half-reasoning *Elephant* is made to associate with the most discordant and stupid . . . *Sloths, Ant-eaters,* and *Armadillos*" as because "my vanity will not suffer me to rank mankind with *Apes, Monkies, Maucaucos,* and *Bats.*" But Pennant went on to mitigate his harsh assessment, acknowledging that Linnaeus's classification of fishes, insects, and shelled animals merited praise and that, especially, "he hath, in all classes, given philosophy a new language."[5]

And more than a generation later, even the nationalistic John Fleming paused in the midst of denouncing the "greatly overrated" Swede to make a grudging nomenclatural concession: "Linnaeus was not, it is true, without much merit, in rendering trivial names [that is, technical species names] popular."[6] Fleming used the word *popular* in a rather restricted sense, referring only to the acceptance of "trivial names" within the community of naturalists. Indeed, wider acceptance might have made them less popular among Fleming's own cohort, for at the same time that it provided a newly systematic means of referring to animals and plants, Linnaean terminology also offered a newly definitive means of discriminating between the zoological knowledge of specialists and the implicitly less significant and reliable information about animals that was broadly available to ordinary Britons. It redefined a reservoir of traditional lore as the turf of experts.

The Power of the Word

The intensity of this durable admiration for Linnaean nomenclature was in direct proportion to the magnitude of the chaos it was understood to have superseded. At the beginning of the eighteenth century, in the opinion of most subsequent naturalists, the names of animals were at least as likely to hinder zoological investigation as to advance it. The long centuries during which animal lore had been unsystematically collected, either by pedantic bestiarists or by unschooled countryfolk, had left at once too many names and too few. Any familiar or even widely recognizable animal was likely to have accumulated an abundance of synonyms in every European vernacular. This superfluity often made it difficult for naturalists to recognize that they were discussing the same creature. In his retrospective survey, for example, Pennant listed multiple options in English, French, German, and indigenous North American languages for the reindeer. The hippopotamus, which few contemporary Europeans had actually seen, had nevertheless accumulated names in Latin, Greek, French, and Tgao (an African language); its English appellations included river horse, sea horse, behemoth, river paard, and water elephant.[7] On the other hand, a single name might signify several distinct, if similar creatures. Anglophone adventurers were apt to refer to both the jaguars of South America and the leopards of Africa and India as "tigers"; and they were conversely apt to ponder the difficulty of distinguishing the "leopard" from the "panther." Naturalists had occasionally attempted to rise above this vulgar polyphony by coining names in the learned tongue of Latin, but in so doing they simply produced an additional layer of confusion. For example, by the early eighteenth century, the leopard and the panther had, between them, accumulated the Latin denominations of *Panthera, Pardus, Pardalis, Leopardus,* and *Uncia,* some of which were occasionally applied to the cheetah (then usually termed the "hunting leopard" in vernacular English), the jaguar, and various lynxes as well.

This unmanageable profusion of names, with its concomitant blurring of the boundary that distinguished scientific expertise from other modes of knowledge, was a constant irritant to working naturalists. When Edward Tyson tried to imagine an ideal zoological reference work, his first desideratum was that the entries "may contain The Names Synonyma's of our own or other Nations both Antient and Modern."[8] And if the multiple and inadequately delimited names of well-known creatures could impede the progress and undermine the dignity of research, the effect of such problematical nomenclature on the study of unfamiliar animals was still more pronounced. The propensity of early explorers to name American creatures after those of the Old World produced widely lamented consequences; indeed, the surviving names of the mountain lion or panther and robin still cause transatlantic confusion. As one English interpreter of Buffon put it, "to avoid falling into perpetual errors, it is necessary to distinguish carefully what belongs to the one continent from what belongs to the other."[9] Similarly, seventeenth-century European observers tended to label the indigenous fauna of southern Africa with names borrowed from homelier creatures, a practice which implicitly discouraged those who relied on their accounts from perceiving these animals as very distinctive or unusual.[10] The first Dutch settlers used the suffix *bok,* their term for male goat and a cognate of one English term for male deer, to designate the dazzling variety of antelopes that they found in the Cape region.

Despite the condescension they predictably inspired, the confusing records left by such unsophisticated observers were not invariably superseded by the more precise and systematic efforts of subsequent investigators. Because of the enormous changes in faunal distribution that could result from European colonization, or even contact, more enlightened naturalists often had to rely on whatever cryptic accounts were available for insight into the pre-existing state of colonial nature. Such accounts would inevitably also have served as reminders of the bad old days, when nomenclature had proliferated without discipline. In 1819 Charles Hamilton Smith lamented to fellow members of the Linnean Society that the unfortunate vernacular naming practices of the original Dutch settlers of New Jersey made it difficult to decide whether a particular species of antelope had ever lived in that state.[11] He spoke, however, with resignation, even complacency, rather than anguish. By then, his tone implied, this misleading and unsystematic nomenclature could be understood as merely a kind of pentimento, a ghostly reminder of long-suppressed disorder.

A more persistent nonspecialist source of zoological information inspired similarly mixed responses. Indigenous peoples had unmatchable access to the fauna among which they lived. Despite their alleged indifference to systematic natural history, therefore, every scrap of data they offered appeared at least worthy of scrutiny. As John Hunter put it, "even the name given by the natives should be known if possible; for a name to a naturalist should mean nothing but that to which it is annexed."[12] Although it was potentially illuminating, however, such evidence was also considered unreliable and difficult to interpret. It posed problems similar to those embedded in the accounts of naive Europeans. Indeed, sometimes these two kinds of sources were conflated, as if the shared lack of zoological expertise could unite colonial peoples otherwise divided by race and nation. William Burchell, when listing his South African specimens, consistently opposed a miscellaneous series of vernacular alternatives to a single definitive latinate binomial. For example, the *Antilope oreas* was "The *Eland* of the Dutch Colonists . . . Knna in the language of the Hottentots, and Pō-hu in that of the Bachapins."[13] Barely a year and a half after Smith had regretted the terminological slackness of the seventeenth-century Dutch farmers, the same Linnean Society audience heard Stamford Raffles lodge a parallel complaint about the imprecise characterizations of Sumatran forest animals—even quite large and distinctive ones, such as tapirs and rhinoceroses—that could be extracted from native reports: "a perfect consistency or uniformity of nomenclature cannot be expected, and it is not always easy to reconcile the synonymy."[14] And later still, George Vasey cautioned with respect to the wild bovines of India, "when we hear one animal called Gayal and another Gyall, we are not, *on that account merely,* to set them down as of the same species."[15]

The evolution of the term *nondescript* also indicated zoologists' increasing confidence in the power of nomenclature to define the sphere controlled by specialist knowledge. In the seventeenth and eighteenth centuries it was a rather neutral term for species that had not yet been described and labeled by naturalists. Early in the nineteenth century William Bullock used it in this sense, when he referred to several large but obscure fossil bones on display in his museum as the remains of "stupendous nondescripts." When Bullock praised his "Incognita" as "the surprise of the enlightened naturalist, and the admiration of the classical scholar," however, this restricted characterization of the audience for "nondescripts" was already becoming outdated.[16] Among specialists, indeed, lack of a name could be more of a reproach than an attraction. Nondescripts were apt rather to appeal to casual curiosity-seekers with no stake in eliminating the mysteries of nature, or even with a preference for keeping deep waters uncharted. They were increasingly likely to figure, not in learned zoological discussions, even of unnamed creatures, but in the posters and handbills advertising sideshows and menageries. In 1817, for example, four wapiti on display in a mews near Charing Cross were (somewhat misleadingly) described as belonging to "an extraordinary species of NEW AND NON-DESCRIPT DEER, lately arrived from N. America."[17] More commonplace creatures similarly borrowed heightened charisma from this epithet. Later in the century the bearded Julia Pastrana was billed as "the Nondescript," and the same term was applied to "an ape-like creature," actually an Indian boy who had been raised by wolves.[18] The credulity implicit in the search for crowd-pleasing nondescripts rapidly became the target of satire or parody, most famously in the stuffed creature whose very hairy but distinctly human features graced the frontispiece of Charles Waterton's *Wanderings in South America* (1825). A taxidermical innovator as well as a zoological eccentric, Waterton had fashioned his "Nondescript" out of the hindquarters of a red howler monkey.[19]

Nineteenth-century naturalists felt that they could afford to disparage the nomenclatural chaos that had been pushed to the temporal and geographical peripheries—and to use command of nomenclature to separate initiates from non-initiates—because of their faith in the comprehensiveness and order of the system of naming attributed to Linnaeus. In this system, every organism was designated by a single name, consisting of two latinate words, the first indicating its genus and the second its species. Further refinement or differentiation could be provided by a third term, indicating what could be understood as subspecies or race or variety. But this form of faith, like others, owed as much to will as to demonstration. Useful though it incontestably was, Linnaean nomenclature was far from flawless. One shortcoming was that, despite its celebrated novelty, it could be confused with discarded pre-Enlightenment terminology. Because of their classical form, Linnaean terms often resembled those employed by earlier naturalists and bestiarists, whose Latin binomials and trinomials were, however, simply abbreviated or economical descriptions, not unique and systematically gen-

Charles Waterton's "Nondescript."

erated designations.[20] Linnaeus himself had exacerbated this difficulty, by loading some of his names with an onerous burden of description in addition to mere designation. Further, Linnaean nomenclature begged many questions of importance to eighteenth- and nineteenth-century naturalists, such as whether the species or the genus was the fundamental unit of classification and, still more difficult and fundamental, how individual species were to be recognized and circumscribed.

But the system could be used to produce unlimited numbers of unique names, and its very lacunae made it appropriable by zoologists of almost any taxonomic persuasion. In fact, the consensus made possible by the severance of Linnaean nomenclature from any particular system of classification itself encouraged the further broadening of that consensus. For example, William Lawrence, who had deep reservations about Linnaeus, nevertheless felt obliged to offer an explanation of why some of his own "terms . . . differ occasionally from those of the Linnean system, which has been hitherto chiefly followed in this country." He emphasized that such divergences should be kept to a minimum because "the introduction of new . . . names must . . . create difficulty and confusions."[21] By the early nineteenth century, when Linnaean nomenclature was understood to be more or less universally accepted among British naturalists, it had become primarily a technique for the coding and retrieval of information.[22]

As with classification in general, however, there was many a slip between the abstractions of nomenclatural theory and the concretions of nomenclatural practice. Even reduced to a method of indexing, Linnaean nomenclature was far from simple. It required that a complex and ambiguous set of rules be applied to raw material that could be characterized in the same terms, by naturalists who themselves varied widely in culture, disciplinary background, and personal commitment. Perhaps it was not surprising that their nomenclatural applications showed equivalent divergences as well as inconvenient overlaps. Again, Linnaeus himself could falter in this regard. His admirer George Shaw apologized for the fact that, because of a "confusion and misapplication of synonyms," the variegated baboon ("*Simia mormon*") had been "confounded with one really different, though very much resembling it," noting that "in so extensive a work as . . . the Systema Naturae" such lapses were "almost unavoidable."[23]

Subsequent nomenclators proved no better at avoiding them, and once again designations proliferated, albeit in the prescribed form of latinate binomials and trinomials. An early-nineteenth-century owner of the 1793 edition of Thomas Pennant's *History of Quadrupeds* repeatedly found the list of latinate synonyms that began each entry to be insufficient; he was frequently obliged to pencil additional designations in the margins.[24] In 1830, the museum of the Zoological Society of London was criticized for the "barbarous assemblage of names, as if to describe all the mongrels in creation," with which a single wild goat was labeled.[25] And in 1896, looking back on more than a century of post-Linnaean primate nomenclature, Henry O. Forbes abjured the attempt to "write a synonymy of the species of Monkeys"—that is, to collect all the names by which naturalists had denominated each species. Not only was the relevant information "scattered over many, often obscure, periodicals," but he feared that the consequence of assembling it might be "to introduce a great deal of confusion."[26]

To a certain extent these gaps between the promise of Linnaean nomenclature and the results it actually delivered reflected technical problems incident to the work of natural history. To ensure that an apparently new species had not previously been discovered, described, and named by someone else, it was necessary, then as now, to search the literature. Networks of transportation and communication were constantly improving, but not fast enough to guarantee that naturalists would be able to locate and examine all potentially relevant reports—buried as they might be in the proceedings of obscure societies, published in foreign languages. Even if a possible precursor emerged in the printed record, it might be difficult to establish whether the two animals in fact belonged to the same species. A definitive judgment would require the comparison of both specimens, which might be irrevocably separated by geography or by condition of preservation, even assuming that they represented the same sex, age, or life phase. Few naturalists had the resources of time, money, and prestige to match the efforts made by Charles Darwin as he worked on his monograph about barnacles: for several years in the 1840s his house was filled with smelly specimens loaned by a global array of scientists, private collectors, and curators.[27]

Even if these formidable difficulties could be overcome, nomenclature might proliferate as a result of what were recognized as legitimate differences of zoological theory or practice. Then as now, taxonomists were divided into "splitters," who were inclined to recognize species and higher taxa on the basis of relatively slight differences, and "lumpers," who advocated a higher threshold for separation.[28] For example, although Linnaeus had established a single genus, *Equus,* to accommodate the horse and its close relatives, many subsequent naturalists wished to acknowledge subdivisions within this group by creating separate genera for asses *(Asinus)* and for zebras *(Hippotigris)*. In criticizing this practice, one committed late Victorian lumper suggested that nomenclatural or linguistic considerations should take precedence over those of anatomy, arguing that, "the great inconvenience of altering the limits of genera is that, as the name of the genus is part of the name by which . . . the animal is designated in scientific works in all languages, every change in the limits of a genus involves some of those endless changes in names which are among the greatest causes of embarrassment in the study of zoology in modern times."[29] This pragmatic argument had some force in the case of equines, for the parties to the nomenclatural controversy agreed about the relationships that the species bore to one another; they differed only about the proper representation of those relationships. But in cases where relationships themselves were moot or contested, shifts in nomenclature were unavoidable. When Thomas Hardwicke discovered a small new carnivore in the Himalayas, the unusual dentition of which placed it outside any of the recognized genera, he felt reluctant to coin a name unlikely to be permanent.[30] And when the balance of zoological opinion shifted to agree with William Jardine that the musk ox, conventionally denominated as *Bos moschatus* and therefore a member of the genus that included ordinary cattle, "may perhaps have an appropriate station as an intermediate form connecting this division with the sheep," there was no avoiding the creation of an additional genus. Its name, *Ovibos*, memorialized its newly liminal position.[31]

Setting Things Right

Nineteenth-century naturalists were, of course, perfectly aware of these problems, which they regularly lamented at the same time, if not in the same paradoxical breath, that they celebrated the transformation in their discipline wrought by the introduction of binomial nomenclature.[32] In 1833 a contributor to the *Field Naturalist* "regretted that . . . the language of zoology and botany is necessarily changing. And what is the consequence? we are overburdened with synonymes, . . . [which] create as much, if not more, confusion than did the provincial terms, in the absence of scientific nomenclature."[33] Nor was their reaction to this oddly intractable situation limited to lamentation. The 1840s saw the beginning of a sustained effort at reform on the part of establishment British zoologists. At the 1841 meeting of the British Association, a committee with a small but distinguished membership was charged "to draw up a series of rules with a view to establishing a nomenclature of Zoology on an uniform and permanent basis."[34] The committee drafted a "Proposed Plan" that was circulated to a long list of British naturalists and a short list of foreigners; a "Proposed Report of the Committee on Zoological Nomenclature," modified in response to their comments, was printed in 1842 and the rules it suggested were adopted by the British Association.[35]

These labors received a good deal of private and public praise. More tangible results were, however, thinner on the ground, and in 1865, after several disappointing decades, the British Association was moved to readopt the proposal, only slightly modified by the few surviving members of the original committee.[36] Again, the positive impact on zoological practice was difficult to discern. In 1874, Alfred Russel Wallace observed that although "zoologists and botanists universally adopt what is termed the binomial system of nomenclature invented by Linnaeus," nevertheless "one of the first requisites of a good system of nomenclature—that the same object shall always be known by the same name—has been lost."[37] In consequence, as William Henry Flower asserted in his 1878 presidential address to the British Association's Zoology and Botany Section, "all beginners are puzzled and often repelled by the confused state of zoological nomenclature."[38]

From the straightforward perspective of efficiency and utility—the values that were ordinarily invoked in discussions of nomenclature—the manifest reluctance of the Victorian zoological community to accept what was generally acknowledged to be a very sensible set of suggestions might be difficult to explain. After all, the confusing status quo required specialists to waste valuable time on what were essentially clerical labors; and even relatively broad-ranging nomenclatural debate had long ceased to help controversialists onto the scientific fast track. On the contrary, it was "well-known to everyone . . . that no department requires so much patience and impartiality, and shows so little apparent result in comparison with the time and labour bestowed."[39] As Alfred Newton, the first professor of zoology at Cambridge, noted in 1879, when he urged his colleagues to adopt the still-orphaned British Association rules, "nomenclature is so trifling an adjunct to zoology that no true student of the science can fail to grudge the time which he is . . . compelled to bestow upon it, or ought to be ungrateful to those who have expended their toil in preparing some rules for his guidance through the intricate maze of synonyms that . . . enfolds almost every object with which he has to deal."[40]

Similarly, given the apparently mundane and pragmatic nature of the issues surrounding scientific nomenclature, it could be difficult to account for the tone of anxiety and passion that frequently crept into learned discussions of it. The initial proposal circulated by the British Association committee in 1841 characterized nomenclatural irregularity as an "evil," the result of "neglect and corruption"; it referred to Buffon's practice of christening new species only in the vernacular and not with latinate binomials as "vicious."[41] In his response to the draft proposal, W. J. Broderip, a successful lawyer as well as a respected naturalist, implicitly acknowledged the volatility of the topic when he warned against using words like *Parliament* or *legislation*, which might give "the appearance of dictation" and thus "excite ridicule."[42] Such language suggested that more was at stake in establishing uniform and consistent zoological nomenclature than the elaboration of a merely technical order.

Indeed, the naturalists who drafted the original British Association proposal began by dismissing technical sources of confusion—"those diversities which arise from the various methods of classification adopted by different authors, and which are unavoidable in the present state of our knowledge"—as of secondary concern.[43] Instead, they focused their attention on discrepancies that arose from extra-disciplinary causes. Challenges to the intellectual authority of elite British naturalists were conflated with challenges mounted on other grounds, more clearly rooted in human nature and therefore more vulnerable to policing. Nomenclature became a medium upon which a variety of frailties and lapses and antagonisms could be inscribed, as well, inevitably, as the representative or symbol of those alternative behaviors and commitments. An energetically enforced standard of nomenclatural propriety would embody and reinforce hierarchical order both inside the zoological community and in the larger society to which its members also belonged; at the same time it would identify inappropriate or troublesome colleagues. Consequently, the errors and eccentricities in nomenclature that attracted the most severe and protracted criticism from the British Association committee were those that most clearly associated their perpetrators with groups considered obnoxious for political or cultural or social reasons.

Some of the most provocative challenges were mounted from abroad. In an era of intense international military and political rivalry, scientific claims could be conflated with those of the polity in general; the clash of soldiers and diplomats had its analogue in the nomenclatural activities of zoologists. Naming constituted a strong, if metaphoric, claim to possession, not only of the newly christened species, but by implication of its native territory; conversely, territorial claims were easier to question in learned journals than on the battlefield. Stamford Raffles once found himself in the unhappy position of having to dismiss "two French gentleman who [had] appeared qualified" to help him with the preservation and description of the many specimens he had collected during his colonial service in Southeast Asia, lest, as a result of what he called their "private and national views," "all the result of all my endeavours . . . be carried to a foreign country." What he feared was the integration of his specimens into a Gallicized nomenclature—which he characterized as "speculative and deficient in the kind of information required"—and their consequent loss not only to himself but to his nation.[44] The ornithologist John Gould attempted this maneuver in reverse when he named a South American species *Rhea darwinii,* even though it had already been otherwise christened by Alcide d'Orbigny.[45] Thus, ironically, the Linnaean terminology originally designed to serve the supranational scientific community—and for that reason, among others, couched in latinate forms that recalled the universal language of medieval and Renaissance learning—had come to replicate the separation of rival national cultures. Later in the nineteenth century, such separatism in the clothing of universality drove Edwin Ray Lankester, whose distinguished career featured professorships at Oxford and London as well the directorship of the British Museum (Natural History), to suggest that his colleagues abandon their internationalist aspirations and content themselves with imposing uniform terminology at home. He urged the introduction of "a series of terms distinctly English in their etymology, which would be accepted as authoritative and used throughout the country."[46]

The prominence of political concerns, as well as the fact that, like Raffles, many naturalists also participated in the imperial enterprise as government administrators, military officers, or explorers, meant that the first nomenclatural lapses singled out for criticism by the British Association committee were those committed by foreign naturalists. The published report of 1842 lamented that "the commonwealth of science is becoming daily divided into independent states . . . If an English zoologist . . . visits the museums and converses with the professors of France, he finds that their *scientific* language is no less foreign . . . than their *vernacular.*"[47] In making this complaint, the committee followed a trail blazed by earlier British critics, who had identified flaws in the Gallic national character that might account for this willful and uncooperative divergence: the French "rage for innovation" and preference for "forever subdividing where the great aim should be to combine."[48] It was significant that France, Britain's most serious geopolitical competitor, figured as the primary locus of the linguistic "despair" experienced by traveling British naturalists, with Germany and Russia mentioned only as afterthoughts. Perhaps even Buffon's practice would not have seemed so vicious if he had abjured latinate nomenclature for that of some other vernacular. Diplomatic considerations also shaped the strategic planning of the British Association committee; for example, the explorer John Richardson prophetically cautioned, "the main difficulty will be in gaining the hearty assent of the European naturalists to the proposed plan."[49] These forebodings were confirmed as the scientific establishments of other nations responded with reciprocal patriotism. The immediate effect of the British Association's initiative was to exacerbate the international Babel of nomenclatures, by inspiring naturalists throughout Europe and North America to promulgate their own competing plans.[50]

But the foreign menace was not the only one with which the British Association committee felt it had to contend. In the view of the establishment naturalists who composed it, the terminological practices of some fellow citizens presented similarly grave challenges to the order and hierarchy that binomial nomenclature had been designed to embody. One of the most disturbing of these, castigated as an "evil" in the committee's initial report, was "the practice of gratifying individual vanity by attempting on the most frivolous pretexts to cancel the terms established by original discoverers, and to substitute new and unauthorized nomenclature in their place."[51] The identification of a new species was, after all, an important achievement for anyone seriously interested in the natural world. The sportsman William Cornwallis Harris, after a hunting expedition that yielded over "four hundred head" of large animals, including "every known species of game quadruped in Southern Africa," claimed his efforts had been "crowned by a truly splendid addition to the catalogue of Mammalia"—a previously unknown species of antelope.[52] For zoologists, the discovery and, especially, the description of a new species provided solider rewards, including the chance to choose the organism's latinate binomial, which would henceforth be formally cited with the name of the namer appended in parentheses. By this act, individuals claimed intellectual possession of the animal, in parallel with their government's territorial claim. The lack of such an accomplishment could diminish the luster of a scientific reputation, as when Richard Lydekker qualified his assessment of William Flower's career because "Flower never named a new species of animal, nor . . . did he ever propose a new generic term."[53]

Although scientific etiquette forbade describers of new species to commemorate themselves (an attempt to enforce refined self-effacement that was in any case routinely compromised by their tendency to name species after their employers or patrons),[54] this parenthetical convention provided a strong incitement to naturalists who might inappropriately value glory over truth. The most frequently deplored means of accomplishing this end were excessive splitting—the carving up of previously recognized species so as to produce new entities that would require new names—and the too careless assumption that a new specimen represented a new species. In 1878, *Nature* castigated such slack and ignorant colleagues as caring "very little to know what others are doing."[55] Sins of commission provoked severer censure than those of omission. As one purist irascibly put it, "nomenclators who take a *pride* in the manufacturing of names . . . fight among themselves as to which . . . shall have the *honour* of appending his name to them, hoping that by thrusting a Jack Scroggins-of-a-name into notice, it will be handed down to posterity."[56]

The terms in which elite naturalists denounced such weaknesses often evoked the getting and spending of tradesmen, thus implying social condescension as well as scientific disapproval. In several letters to H. E. Strickland, who chaired the British Association committee, Darwin disparaged those colleagues in need of suppression as mere "species-mongers" who wanted to "have their vanity tickled" and were therefore responsible for a "*vast* amount of bad work." In response, Strickland compared those who "hurry to be *first* in the species-market" to "Covent-gardeners with their green peas, a guinea a pint, at Xmas, and unripe strawberries, a shilling a dozen, in April." More harshly, the pseudonymous Solitarius criticized "those whose paltry conceited minds are gratified at the idea of having obtained a little celebrity for themselves, by the shortest and easiest method."[57]

The suggestion of vulgarity latent in these characterizations reemerged in the rules proposed as specific correctives for egotistical excesses, rules that perhaps surprisingly stressed the importance of philological and grammatical correctness in the creation of zoological nomenclature. Since correctness was understood in terms of a command of classical languages normally acquired only through elite education (if then), it also served to distinguish zoologists with privileged backgrounds from those whose expertise had been acquired in less genteel academies.[58] It was asserted "the *best* zoological names are . . . derived from . . . Latin or Greek," and namers were warned against designations that revealed a misunderstanding or half-understanding of classical texts, such as referring to an ancient name for a different animal or a mythological figure that had no relation to the character of the animal being named.[59] Indeed, as Charles Lyell pointed out in *Principles of Geology,* terms coined or adapted for the general scientific vocabulary could also betray lack of refinement, if the inclusion of "foreign diphthongs, barbarous terminations, and Latin plurals" signaled insensitivity to the subtleties of English as well as of other tongues.[60] That such strictures reflected a desire to establish a binary taxonomy of Victorian zoologists, rather than any more generalized respect for traditional scholarship, was suggested by the British Association committee's warning to nomenclators inclined to delve too deeply into Aristotle, Pliny, and other figures from the prehistory of zoology. Such propensities might result in "our zoological studies . . . [being] frittered away amid the refinements of classical learning."[61]

Further, whole categories of names were repeatedly banned on the grounds that they revealed lack of taste: for example, nonsense names, names made up of fragments of two different words, hybrid names that combined elements of two languages (the inability or disinclination to distinguish between Greek and Latin roots was often also a telltale indication of extracanonical education), and names that instead of commemorating "persons of eminence as scientific zoologists," celebrated "persons of no scientific reputation, as curiosity dealers . . . , Peruvian priestesses . . . , or Hottentots."[62] A paleontologist primarily interested in the remains of extinct sea monsters characterized the latter transgression as "injurious to the dignity of Science, and the Taste of the Age in which we live."[63] And the entomologist Francis P. Pascoe, who would have read the British Association committee's original report as a young naturalist, vociferously restated its complaint in old age, protesting in the introduction to a manual of zoological taxonomy "against the barbarous and other objectionable names (sometimes at variance with good taste and even with decency) that have been introduced into science—such, for example, as Battyghur, Butzkopf, Agamachtschich, Know-nothing, Stuff, Jehovah, Cherubim, or such idiotic names, or rather sounds, as Toi-toi, Sing-sing, Gui, Yama-mai." Having thus exemplified lapses in refinement, he left his readers to imagine still more heinous offenses, concluding that "indecent names need not be further alluded to."[64]

Nor did déclassé naturalists pose the only domestic challenge to the authority represented by a uniform, authoritative, and exclusive code of zoological nomenclature. The large general audience for zoological curiosities had little interest in the terminological hairsplitting of nomenclatural purists—or, for that matter, in any binomial designations. As James Edward Smith somewhat implausibly reflected in 1819, if Linnaeus and his contemporaries had imagined "that their science would ever be the amusing pursuit of the young, the elegant, and the refined," they might not have used so much Latin.[65] From this perspective, differences within institutionalized science were less important than the difference between that self-defined elite, which claimed authority over a certain sphere of knowledge, and others interested in the same material, who were apt to resent both the claims themselves and the confident, even overbearing tone in which they were often made. Resentment inevitably crystallized around the provocatively arcane nomenclature of zoology. Thus John Ruskin advocated the replacement of latinate binomials for birds with "the simplest and most descriptive" English nomenclature, as part of an explicit attempt to reclaim natural history for nonspecialists. He made some attempt to argue in the terms of science, claiming that the need to incorporate new and yet-to-be-discovered information made it "one of the most absurd weaknesses of modern naturalists to imagine that *any* presently invented nomenclature can stand, even were it adopted by the consent of nations, instead of the conceit of individuals." But his strongest objection was more fundamental: "a time must come when English fathers and mothers will wish their children to learn English again, and to speak it for all scholarly purposes; and, if they use, instead, Greek or Latin, to use them only that they may be understood by Greeks or Latins; and not that they may mystify the illiterate many of their own land."[66]

In addition to providing a target for criticism, zoological nomenclature could be subverted to express disapproval and even ridicule. Indeed, even people whose sympathies and talents were usually at the disposal of science might be tempted to poke gentle fun at the pretentiousness implicit in an enterprise that relentlessly reclothed familiar objects in impenetrable new names. The humorist Edward Lear, a distinguished zoological illustrator, produced a "Nonsense Botany," which featured such plants as the *Encoopia Chickabiddia,* the *Tickia Orologica,* the *Washtubbia Circularis,* the *Plumbunnia Nutritiosa,* and the *Manypeeplia Upsidownia.*[67] In Charles Kingsley's *The Water Babies,* Professor Ptthmllnsprts desired to capture the amphibious hero in a bucket, in order to study him and designate him "*Hydrotecnon Ptthmllnsprtsianum,* or some other long name like that, for they are forced to call everything by long names now, for they have used up all the short ones."[68] With greater hostility, *Punch* occasionally made pseudo-scientific nomenclature the metaphor for the love of obfuscation shared by scientists and other scholarly specialists, referring to the "Clamour-making Cat *(Felis catterwaulans),* which is well known to all Londoners," the "*Felis omnivora,* or Common Lodging-House-Keeper's Cat," the "Learned British Pig *(Porcus Sapiens Britannicus),*" and the "Rum Shrub *(Shrubbus Curiosus).*"[69] The silly underbelly of zoological nomenclature could be obvious even to the least sophisticated of its victims. In Elizabeth Gaskell's novel *Mary Barton,* when a learned if humble naturalist pooh-poohed an unschooled young sailor's claim to have sighted a mermaid, Jack responded by attacking the nomenclature that symbolized the expertise he lacked: "You're one o' them folks as never knows beasts unless they're called out o' their names . . . and [if] I'd ha known it, I'd ha christened poor Jack's mermaid wi' some grand gibberish of a name. *Mermaidicus Jack Harrisensis,* that's just like their new-fangled words."[70]

A more common, and perhaps a stronger, riposte to the claims implicit in the celebration of scientific expertise was to ignore them completely—to indulge without reservation in what the British Association committee referred to as "the vicious taste on the part of the public" for "vernacular appellations."[71] Or, as a writer who signed himself "Bob" more mildly put it in the *Oriental Sporting Magazine,* "I wish that people who write on sporting subjects in the nineteenth century would call the animals . . . by their proper names."[72] The public, however, remained imperturbably an-glophone and insisted on calling a buffalo a buffalo. It was further disposed to follow its own counsel, rather than that of the scientific establishment, even with regard to vernacular appellations, so that it also called a buffalo a bison or even a bonassus. The misprisions of sportsmen seemed particu-larly troublesome, perhaps because they were repeated so often and with such wide publicity. At the end of the nineteenth century, Richard Lydekker

APTENODYTES PENNANTIS, ESQ.

(*A Sketch taken in the Zoological Gardens*).

Nominal satire.

lamented that "although the importance of a correct and uniform nomenclature . . . can scarcely be over-rated," among sporting authors, "the American bison is commonly spoken of . . . as the buffalo," "Indian sportsmen . . . call the Indian gaur, which is closely allied to the true oxen, the bison," "by an unfortunate confusion the term aurochs has been almost universally applied to the bison, although . . . it is really applicable to the wild *Bos taurus*," and "the four-horned antelope is not unfrequently designated . . . as the chinkara, a name which properly belongs to the Indian gazelle."[73]

And hunters were not the only nonspecialists whose alternative nomenclatural predilections, however misguided or depraved, could influence the printed record, and even scientific practice. When financial matters were at stake, zoologists stuck to their terminological guns at their peril. Members of the general public—and even some amateurs of natural history—were apt to regard latinate binomials as "a Torrent of hard Words . . . [a] Parade of indeterminate Sounds," and consequently to avoid linguistic exposure that might prove painful.[74] Several mid-Victorian zoos failed because their directors refused to pitch the exhibits to the general public rather than to naturalists.[75] More accommodatingly, in 1831 an elaborate guidebook to the newly opened Regent's Park zoo assured readers that "the Editor has studiously attempted to avoid" latinate terminology, instead employing "English terms as definite in their meaning."[76] And natural history museums, no matter how learned their primary audience or how august their stature, also attempted to counteract the impression of exclusiveness and inaccessibility produced by latinate labels. In the comparative anatomy collection at the University of Cambridge, "for the convenience of casual visitors . . . the trivial as well as the scientific names have been appended to all the large skeletons, and in many cases to the small ones also"; and "all the specimens" in the Zoological Department of the British Museum were "marked with their popular and their systematic names."[77]

The Sincerest Form of Flattery

Not everyone with a potentially competing stake in the material of natural history resented the increasing hegemony and prestige of scientific nomenclature. One group of lay specialists in animal matters, the breeders of pedigreed domesticated animals, admired this arcane technical terminology in more or less the terms intended by the naturalists who deployed it. Although disinclined to express their appreciation theoretically, these hands-on experts demonstrated it persuasively through their own nomenclatural practice. Voting with their feet, they borrowed the terminology devised for systematic natural history and adapted it to their own rather different ends.

The breeders' appropriation of naturalists' words constituted a striking exception to the ordinary relationship between the two groups. Although the subject matter of eighteenth and nineteenth-century zoology inevitably exhibited a substantial overlap with that of domestic animal husbandry, the literatures of those enterprises intersected much less predictably. Occasional individuals participated in both communities. For example, Joseph Banks, who served as president of the Royal Society, also wrote widely on agricultural matters, promoting such pet projects as the naturalization in Britain of Spanish merino sheep.[78] But, on the whole, the two voluminous discourses evolved in parallel rather than interactively. Naturalists gave domestic animals more than their share of ink in the zoological surveys and overviews that they industriously produced, but they seldom incorporated the practical insights of stockbreeders and pet fanciers. (Charles Darwin was a famous exception to this generalization in his eagerness to profit from the accumulated wisdom of working-class pigeon breeders, as well as of agriculturalists.)[79]

Reciprocally, animal breeders tended to prefer homegrown expertise, loyally adhering to time-hallowed attitudes and practices, even where they conflicted with the most solidly grounded and widely held zoological consensus. They firmly resisted the occasional attempts of scientifically minded colleagues to bring them up to date—like Everett Millais's *Theory and Practice of Rational Breeding*, which urged breeders to stop "trusting to luck" and instead base their decisions on "Darwin's theory . . . the stupendous work of a stupendous mind."[80] John Cossar Ewart, then Regius Professor of Natural History at the University of Edinburgh and, like Banks, a straddler of the boundary between zoology and animal husbandry, suggested in 1899 that the roots of this stubborn imperviousness might stretch deeper than mere indifference: "in England, where there has been . . . in many quarters a dread of scientific methods akin to that of the supernatural, breeders have failed to distinguish between facts and beliefs."[81] *Punch* proffered an alternative explanation when it imagined the ordinary agriculturalist's reception of Darwin's *Variation of Animals and Plants under Domestication*. This work was widely appreciated in the stockbreeding and pet-fancying press, although perhaps more for the simple recognition it awarded their enterprise than for its weight of evidence and argument. Because of the resulting publicity, "Old John Stockwell . . . as good a judge of a beast as any man in Midlandshire . . . borrows the two stout volumes from his clergyman . . . he makes a conscientious effort to master the polysyllabic difficulties, but fails . . . He tries again . . . [finally] he throws Darwin down, is quite laborious with his pipe, empties his tumbler, takes off his boots, and . . . goes up-stairs to bed."[82]

Whatever reservations John Stockwell and his ilk may have harbored about Darwin's vocabulary in general, they showed themselves more favorably inclined toward at least one of its polysyllabic components. The latinate binomials and trinomials of zoological nomenclature surfaced frequently in the literature of animal husbandry. These terms were not, however, generally employed to designate the species of primary concern to breeders. In print, as in everyday conversation, the zoologists' *Bos taurus, Equus caballus, Equus asinus, Ovis aries, Capra hircus, Sus domesticus, Canis familiaris, Felis catus,* and *Oryctolagus cuniculus* were routinely referred to by stockbreeders and pet fanciers as the ox, the horse, the ass, the sheep, the goat, the pig, the dog, the cat, and the rabbit. Instead, animal husbandry experts were apt to reserve Linnaean nomenclature for more subtly differentiated and controversial taxa, where its connotation of scientific authority enhanced not only their prestige but their credibility. A latinate designation could certify the reality of a "breed" or "strain" when more concrete evidence for its existence seemed thin on the ground.

The Case of the Wild White Cattle

The debate that swirled for over a century around what were known as wild white cattle or white park cattle clearly illustrated the argumentative value of latinate terminology, although it was a rather anomalous debate in terms of the predominant preoccupations of animal husbandry. These striking and famous animals—they were celebrated in verse by Walter Scott and commemorated in oils by Edwin Landseer[83]—roamed a few great estates in Scotland and northern England untroubled by the restraints that burdened ordinary domestic beasts. They were left largely to their own devices as far as food and mating were concerned, and they were shot rather than slaughtered when the herds were culled. Like the emparked deer that lived under similar conditions throughout the British countryside, the white cattle were treated in this way because the magnates who owned them considered them to be wild animals. This attributed wildness transfigured them into creatures inherently different from cattle destined for the butcher or the dairy. If the white cattle left their rugged native haunts, it was to grace the display cases of natural history museums or the enclosures of zoos.[84]

The behavior of the white cattle was often adduced as evidence of their untamed nature. They formed circles to menace intruders before charging;

cows hid their newborn calves in high grass, returning to nurse them only occasionally; when a member of the herd was seriously wounded or ill, the others turned on it and gored it to death. But their most telling claim to wildness was historical or genealogical. They were described by admirers as genetically distinct from other British bovines, the ignoble descendants of domesticated animals imported by the various human conquerors of the island, because their ancestors had been the never-tamed aboriginal wild cattle, or aurochs, that had ruled the primeval forests of Europe. (The last continental aurochs had allegedly been killed in Poland in 1627.)[85] At the end of the eighteenth century, Thomas Bewick, relying heavily on the account offered by George Culley in *Observations of Livestock* (1786), characterized the herds emparked at Chillingham (Northumberland), Chartley (Staffordshire), and a few other ancient estates as survivors of "a very singular species of wild cattle in this country, which is now nearly extinct."[86] Almost a century later, in his introduction to *The Wild White Cattle of Great Britain,* John Storer asserted that "there . . . have existed in this country from the earliest historic times, herds of White Cattle, perfectly distinct, and of a different breed from its ordinary domestic races. Some of these herds seem to have been always wild."[87] His contemporary James Harting claimed

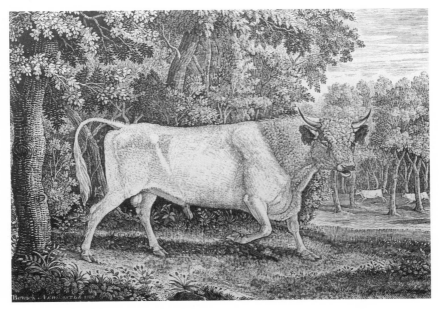

A mighty Chillingham bull, as engraved by Thomas Bewick.

that "the few scattered herds of so-called Wild White Cattle which still exist in parks in England and Scotland may be said to form a connecting link . . . between the wild animals which have become extinct in this country within historic times, and those which may still be classed amongst our *ferae naturae*."[88] When the Earl Ferrers offered the Chartley cattle for sale in 1904, the advertising circular described them as "the lineal descendants of the original British Wild Ox."[89]

The distinguished, often aristocratic, landowners over whose estates the cattle wandered cherished these claims to wildness, aboriginality, and pure descent; they often regarded the animals as family mascots or totems. They were seconded in these feelings by many humbler compatriots, to whom the white cattle appeared as "ancient Britons," representatives of the historical dignity, nobility, and ethnic isolation of their country.[90] To perform this symbolic function, the emparked herds had to be as distinct as possible from "the quiet, servile steer [which] is probably as unlike the original wild cattle of this country, as the English gentleman of the present day is unlike the rude baron of the age of King John."[91] But there was a large element of myth-making or wishful thinking in these assertions of ancient heritage and contemporary wildness, and not everyone was equally transported by nationalistic passion. Less sentimental observers claimed that the vaunted

The king of the Chartley herd.

trappings of uncontaminated wildness reflected human meddling rather than the untrammeled workings of nature.

It was asserted, for example, that even the white color of the herds—an obvious symbol of purity and also, because it was unusual among British domestic breeds, of reproductive isolation—was maintained by killing the off-color (usually black) calves that were frequently produced. The apparent wildness of the white cattle could analogously be explained as the consequence of their style of life; similar behavior would emerge in any group of domestic bovines left to their own devices for a generation or two. Skeptics further pointed out that what looked like wildness to eyes accustomed to the disciplined countryside of the eighteenth and nineteenth centuries had been the standard condition of medieval domestic cattle.[92] In this view, the emparked herds were merely the froward, if interesting, remnants of ordinary cattle that had been liberated at some period in the unfathomable past: possibly feral (that is, having reverted from an earlier state of domestication), but certainly not aboriginally wild.

In the heat of combat, advocates of both positions enlisted the formal authority of zoological nomenclature to buttress their arguments. The advocates of ancient isolation and uncompromised wildness were apt to project the distinction between the emparked herds and other cattle into the past, and to assign the remote ancestors of the white park cattle to a different species than the forebears of their domesticated relatives. Numerous bovines, domesticated and otherwise, had inhabited northwestern Europe in prehistoric and early historic times. In the nineteenth century their skeletal remains were widely available for examination, and many experts assigned them to two or three separate species, defined by the size and shape of their skulls.[93] The largest bones, which belonged, by implication, to the wildest and most powerful animals, were assigned to the aurochs and called *Bos primigenius*. Since the process of domestication frequently resulted in reduced size, remains of smaller animals were designated *Bos longifrons* and *Bos frontosus*.

At the end of the nineteenth century Jacob Wilson boldly, if somewhat nostalgically, asserted that "our British cattle . . . are derived from two species: *Bos Longifrons* and *Bos Urus* or *Primigenius*" and "that the wild cattle of Chillingham are closely allied to the Urus is beyond dispute."[94] But not everyone who sympathized with this position embraced it with such confidence. Although James Harting included a chapter on the "wild white cattle" in a book entitled *British Animals Extinct within Historic Times,* he hedged about "whether . . . [they were] descended . . . from the aboriginal wild breed of the British forests—the Urus of Caesar *(Bos primigenius)*," taking refuge behind "the weight of scientific opinion . . . [which] seems to favour the view that these wild white cattle were descended from the Urus, either by direct descent through wild animals from the wild bull, or less directly through domesticated cattle deriving their blood principally from him." He did, however, firmly distinguish between the urus, for which there was no evidence of early domestication in Britain, and "*Bos longifrons . . .* [which] was everywhere subjugated and used by man."[95] A compromise position, which endorsed the distinctive descent of the white cattle but not their persistent wildness, could assign them forebears distinct from those of other cattle, but not the ferocious aurochs or urus: "Our . . . Chillingham cattle may . . . be descended from *Bos longifrons . . .* Our indigenous kyloes and runts . . . we may assign, perhaps, to *B. frontosus*."[96]

Some believers urged an ancestry still more separated from that of ordinary domestic *Bos taurus.* The author of a mid-nineteenth-century agricultural manual suggested that the aboriginal British wild cattle "still preserved in their native purity . . . in Chillingham park" were descended from a distinct genus *(Bubalus),* that of the "buffaloes" of Asia, Africa, and southern Europe.[97] Thomas Pennant characterized the white cattle as the "*Bisontes jubati* of *Scotland,*" thus connecting them to the European bison or wisent, by then nearly extinct except in protected tracts of the Lithuanian forests.[98] More radical isolationists proposed a genus that would include only the aurochs and its progeny, typically denominating the white herds as *Urus scoticus.*[99] On the other hand, the observation that the emparked cattle interbred enthusiastically and successfully with ordinary British breeds whenever they got a chance sometimes moderated the taxonomic construction of difference, prompting their placement in a separate species of genus *Bos* rather than an independent genus. For example, William Swainson distinguished "Bos Scoticus," which he characterized as "fierce" and "untameable," from the "pre-eminently typical" "Bos Taurus" within his quinary scheme; he viewed their position in the bovine circle as analogous to that of the carnivores in the circle of mammals.[100] And authors on the skeptical end of the continuum could reinforce their position by assigning the white cattle to the same species as shorthorns and Herefords, perhaps acknowledging by terms like "Bos taurus (Linn.) var Scoticus" or "*Bos Taurus. Restricted Variety,*" a difference roughly equivalent to that between ordinary domestic breeds.[101] Or they could emphasize their conflation of the white cattle with more humdrum herds by refusing them any latinate nomenclature at all, simply declaring that "it is . . . highly probable that these animals are the remains of a breed which was formerly kept tame in the farms in many parts of England."[102]

Creative Acts

The constitutive effect of naming and the added prestige of latinate designation did not depend solely on wildness. With reference to more conventional livestock, both binomial nomenclature and the abstract language of
classification (that is, terms like *genus, species,* and *variety*) were routinely
deployed to define, enhance, and celebrate the act of domestication. Eight-
eenth- and nineteenth-century agriculturalists prided themselves on the
extent to which they had "improved" the motley local strains they had found
in British yards and pastures, but the categories in which they consolidated
their advances could seem as much the product of imagination as of the
breeding pen. Pedigreed breeds and sub-breeds were therefore reified in the
language of subspecies or varieties. After all, not only were they, inevitably,
of recent origin, but their characteristic physical qualities could be difficult
to demonstrate. Breeders invoked the prestige and authority of zoological
rhetoric to emphasize the often subtle distinctions that separated one breed
from another, and well-bred from ill-bred animals—that is, to emphasize
and reinforce the refined discriminations on which their success or failure
depended.

The stakes involved in this enterprise, economic as well as intellectual or
technical, were extremely high. The institutional trappings of breed recognition and consolidation—show classes, breed societies, stud books—were
often accompanied by striking and gratifying corollaries. As the editors of
the *Sussex Herd Book* complacently noted, within six years of the appearance
of their first volume, the prediction there expressed "that Sussex Cattle
would come more into favour, has been fully borne out" in increased
demand, and consequently larger profits.[103] The Kennel Club was similarly
founded to promote the interests of dog breeders by holding shows and
field trials, and by ensuring "the proper registration of all exhibited dogs,
with their pedigrees."[104] The lack of such trappings could impede the
recognition of a breed, and consequently the willingness of prospective
customers to pay premium prices for pedigreed animals.

The categorical weakness or lack of focus that breed associations aimed
to correct reflected human failings as well as refractory animal material. All
breeders, whatever the species or variety to which they dedicated their
efforts, strove to produce animals of high and predictable quality. But the
apparent simplicity of this common goal was somewhat compromised by
its vagueness and abstraction. It assumed as given something that, in many
cases, had to be energetically established: the nature of the breed to be
improved and propagated. There was great disagreement among breeders

of every domestic species about which physical attributes characterized an excellent animal. And even if standards could be established, there was further disagreement about which animals should be subjected to them. Breeders concurred in principle that each domestic species should be further divided into subcategories, each with its own aesthetic and functional desiderata; but defining the boundaries and standards for each subcategory inevitably invited further contention. Establishing the existence of a previously unrecognized subcategory—or the validity of a previously disrespected one—could be still more problematic. Since an impressive and authoritative label constituted an implicit argument for the reality and value of the thing labeled, breed partisans frequently attempted to buttress their claims by borrowing the latinate language and the hierarchical categories of zoological taxonomy.

Insubstantial as it might seem, therefore, the argument from nomenclature could prove more powerful than other arguments; and it was certainly more readily available.[105] After all, though pedigree was essentially a genealogical guarantee of purity and excellence, even the best-established breeds rested on uncertain historical foundations. For example, the shorthorn reigned incontestably preeminent among early-nineteenth-century cattle. It was among the first cattle breeds to be judged separately at the annual Smithfield Show, the showcase of British livestock, and in 1822 it was the first breed of livestock animals (as opposed to sporting animals, like the thoroughbred horse or the greyhound) to achieve the dignity of its own stud book, or compendium of individual genealogies.[106] Yet the breed had only coalesced around 1780, with the emergence of the superstar bull Hubback.[107] Although he figured prominently in subsequent pedigrees, Hubback's own origins were obscure. He had begun life as "a calf belonging to a poor man who grazed his cow on the sides of a highway"; it was pure good fortune that a prominent stockbreeder had passed by and perceived his great potential.[108] The roots of thoroughbred horses, the best-bred animals of eighteenth- and nineteenth-century Britain, were not much deeper. All traced their descent from three imported stallions of the late seventeenth and early eighteenth centuries, animals who were, though renowned, semi-anonymous, identified only by their English ownership and their somewhat imprecisely supposed country of origin: the Darley Arabian, the Godolphin Barb, and the Byerley Turk.[109]

The logic that attracted breeders to the language of formal classification might be extended still more boldly and imaginatively. In addition to underwriting or reifying differentiations already incarnated in barn and showyard, zoological terminology might be used to create distinctions that

Cold blood at its worst.
Original caption:
TYPE OF CONVEX COLD-BLOODED HEAD WHICH IS THE ORIGIN
OF COARSE HEADS, INCLUDING BARB AND SPANISH
Betrays slow blood

were desired but not yet achieved. For example, James Long acknowledged in 1886 that, as a result of the difficulty of distinguishing one kind of swine from another, "it is questionable whether agricultural stock of any description has been subjected to such extraordinary classification as the pig"; he nevertheless hopefully asserted that "the Small White may . . . be termed the 'fancy' breed of the pig *genus*."[110] An equivalent process of projection or wishful thinking could also be applied to human beings, who were widely recognized to share many physical and behavioral tendencies with other domesticated animals. For example, Robert Knox asserted that "men are of various Races; call them Species, if you will; call them permanent Varieties; it matters not . . . in human history race is everything."[111] Reversing the analogy, Judith Neville Lytton separated horses "into two distinct types, the Northern and the Southern, which roughly follow the human types of North and South," labeling them "*E.* [*Equus*] *Frigidus*" and "*E. Ardens*."[112]

A further attraction of taxonomic jargon was that the term *breed* itself had become unstable. Originally what defined a breed was shared provenance and shared function, along with some degree of physical resemblance. By the middle of the eighteenth century, however, at least among the most enlightened stockbreeders, the term was generally applied in a narrower sense. As one agriculturalist dedicated to improvement put it, certain "varieties have been usually distinguished among farmers by the appellation of different *breeds;* as they have supposed that their distinguishing qualities are, at least, in a certain degree, transmissible to their descendants."[113] But retrograde usages persisted, even within the community of elite breeders, especially as a way of disparaging unimproved animals. In 1794, for example, a surveyor commissioned by the Board of Agriculture referred to the "original black breed" of Carmarthenshire cattle as "ill shaped and unprofitable to the pail," and to the local "breed of swine—a narrow, short, prick-eared kind."[114]

These general definitional problems were exacerbated by the lack of consensus about the basis for subdividing particular species. The dog was the most emotionally versatile and physically plastic of common domesticates, "more tractable than any other animal, conform[ing] himself to the . . . habits of life of his master."[115] When dog fanciers attempted to make sense of the dizzying variety of canine types, they often found it necessary to identify some intermediate category between the species and the breed. The principles according to which these categories were defined were both numerous and variable. Perhaps the most commonly cited scheme was based on the work of John Caius, whose *On Englishe Dogges,* originally published in 1576, remained influential for several centuries. It divided dog breeds into three unequal categories: "the most generous kinds," further subdivided into "dogs of chace" (hounds and terriers), "fowlers" (spaniels and setters), and "lapdogs" (small spaniels); "farm dogs" (shepherd's dogs or collies and mastiffs); and "mongrels."[116] According to a still simpler principle of analysis, "we may divide our domestic dogs into three great classes:—1. Whole-coloured dogs . . . 2. Parti-coloured dogs . . . 3. Pure white varieties."[117]

Of the fanciers who sought more sophisticated categories, some paradoxically separated dog breeds into the "wild" (which usually included collies, huskies, and others with some resemblance to wolves) and the "domestic."[118] The resulting groups overlapped significantly with those generated by the more anatomically oriented separation of breeds with "erect pointed ears" from those with "pendant ears."[119] Some Victorian breeders acknowledged Cuvier as their authority for a tripartite division

based on the conformation of the parietal bones of the skull (they could be
convergent, parallel, or divergent), which yielded categories represented by
the greyhound, the setter or spaniel, and the mastiff.[120] And fanciers often
grouped dog breeds on the basis of function, a criterion which could itself
give rise to further disagreement. Thus the strictly utilitarian categories
proposed by Peter Lund Simmonds—"farm dogs . . . hunting dogs . . . and
shooting dogs"—excluded the majority of pets, while the more relaxed
categories of the Kennel Club—"Sporting Dogs" and "Non-Sporting
Dogs"—left no one out.[121]

The incompatibility of these competing vernacular differentiations inevitably threatened to undermine the authority of those who made them. To
shore up the eroding foundations of their expertise, nineteenth-century
fanciers frequently dignified their systems of breed classification with terminological trappings borrowed from the rich zoological literature on domestic dogs, the attractiveness of which was undiminished by the fact that
naturalists were as polyphonic about the interrelationships of dog breeds as
were fanciers. Charles Hamilton Smith attached a latinate binomial to each
domestic breed in the *Dogs* volume of the Naturalist's Library; Edward
Griffith repeated Linnaeus's own labels for domestic dog breeds in his survey
of the carnivora.[122] Surveying the animals of Britain, John Fleming more
modestly labeled each of the "principal races of dogs" with a trinomial or
subspecific tag. The collie was *Canis familiaris pastoralis,* the Newfoundland
C. familiaris amphibius, the bulldog *C. familiaris taurinus,* the mastiff *C. familiaris mastivus,* and even the cur *C. familiaris villaticus.*[123] In its early years,
the Regent's Park zoo displayed caged examples of "many interesting varieties of the *domestic dog*" sporting such labels as "*Canis familiaris,* var.
borealis" (eskimo dog) and "*Canis familiaris,* var. *Australasiae*" (dingo).[124]
Fanciers could thus pick from a rich array of possibilities. For example, in
the introduction to his compendious *Illustrated Book of the Dog,* Vero Shaw
surveyed several centuries of dog breed classification, carefully reproducing
every one of the long-discarded latinate binomials listed by Linnaeus, even
those which, like "Canis Hybridus, or Bastard Pug Dog" and "Canis Vertigus,
Turnspit," referred to animals that his Victorian contemporaries would have
found very difficult to recognize, or even to imagine.[125]

Cattle breeds were fewer and, on the whole, slightly more venerable than
dog breeds; in addition, their characteristics were not quite so volatile, owing
as much to the smaller number and greater homogeneity of breeders as to
the lesser variability of the animals themselves.[126] Nevertheless, in order to
reconfirm the claims to distinctiveness and merit of their favorite breeds,
aficionados of shorthorns, Herefords, and their ilk similarly analyzed and

subdivided domestic cattle. As with dogs, some of the resulting groupings were organized on completely vernacular principles. Such principles might be geographical, as in an eighteenth-century agricultural manual that recommended "your *Angleseys* and *Welsh*" as "a good hardy sort for fatting on barren or midling sort of land."[127] Or they might be based on superficial physical characteristics. For example, some experts claimed that "the cattle of Great Britain . . . may be divided . . . into the following primary sections: Long Horns; . . . Middle Horns; . . . Polled Cattle; . . . Short Horns." Not only were such categories grounded exclusively in practical experience, but their formulators often explicitly denied any relationship with the authority of science. In this way, after propounding his horn-based categorization, W. C. L. Martin stipulated that "the . . . sections do not . . . derive their nomenclature from points of zoological importance."[128]

He may have been protesting too much, however; his use of the word *nomenclature* itself suggested at least a tacit parallelism. And, like dog fanciers, many of his fellow breeders attempted to buttress their favorite subdivisions by explicitly enlisting the categories of zoology. An early-nineteenth-century agricultural pamphleteer noted that "the different breeds or varieties of our domestic animals are, in general, so fixed and incapable of speedy alteration, that they may be deemed comparatively permanent."[129] Relatively forceful and lucid stockbreeders could make direct assertions of analogy, claiming that "the distinction between some [breeds] . . . has scarcely been less than the distinction between that variety and the whole species" or, more sweepingly, that "as most wild species consist of several varieties, so must domestic breeds consist of several strains."[130]

It was therefore a mark of distinction among cattle breeds to be further broken down into named and recognized subdivisions, each smaller category simultaneously reifying itself and validating the one above. The introduction to a late Victorian herd book explained that "TRIBE is used in referring to the whole of the descendants . . . of each originally selected Cow . . . Distinguished collateral Branches of a Tribe . . . are spoken of as Distinct FAMILIES . . . The representatives of such TRIBES and FAMILIES, now existing . . . are classed as forming one GROUP."[131] In the case of shorthorns, an extra layer of differentiation commemorated the competition between the eponymous "sub-breeds" established by master breeders Thomas Booth and Thomas Bates.[132] Most tribes and families of cattle, however, were named for a bovine mother, rather than a human father, to the annoyance of some macho breeders, who feared that it would lead their colleagues "to overestimate the value of the female side of the pedigree."[133] In the late nineteenth century, the "Leading Families" of polled Aberdeen

Angus cattle included the Queen tribe, the Ericas, the Princesses, the Baronesses, and the Emilys; an advertisement for shorthorns from the Prince of Wales's estate at Sandringham offered the Cold Cream, Seedling, and Primula families of the Booth strain as well as the Bates families of the Blanches, the Wild Eyes, the Honeys, and the Sweethearts; and Amos Cruikshank, a latter-day challenger for the mantle of Bates and Booth, had developed his own widely recognized set of strains, including the Venus family, the Lovely family, and the Spicy family.[134]

Subdivisions—and names—of this sort evoked the taxa of natural history without literally borrowing them. "Families" were part of the canon of zoological categories (most commonly, in order of diminishing inclusiveness, kingdom, phylum, class, order, family, genus, and species), and "tribes," while less formally established, had long hovered on the periphery as an option for naturalists who felt constrained by the seven-step hierarchy. Both categories occupied similar stations on the taxonomical scale; indeed, *tribe* could function, like *kind*, as a vernacular alternative for *family*. A zoological tribe, like a zoological family, was composed of several related genera; thus William Swainson included "the genera of the elephant, rhinoceros, megatherium, and hippopotamus" within "the thick-skinned or pachydermatous tribe."[135] In confirmation of the taxonomic importance that naturalists accorded them, most species of common domesticated animals gave their latinate names to both the genera and the families of which they were part. For example, the genus *Bos*, which included domestic cattle and their near relations, belonged to the family Bovidae, along with bison and buffalo, and the genus *Canis*, which included domestic dogs and wolves, belonged to the family Canidae, along with foxes, dholes, and the so-called wild dogs of Africa.

Variations on the Theme

When breeders of domesticated animals applied the taxa and nomenclature developed by naturalists for wild species to their pedigreed strains, the dynamic of appropriation seemed straightforward enough. Assuming that the process of zoological classification had relatively firm intellectual moorings and that its objects had relatively real physical existence, breeders implicitly claimed these qualities for their own shakier enterprise. But they also conjured with the prestige of science in more complex, assertive, and even playful ways. Their adaptation of terms like *family* and *tribe* to designate groups of pedigreed livestock, for example, suggested that they could transform zoological categories, as well as imitate them. After all, the families and tribes of stockbreeders were not at all like those of naturalists. On the contrary, they designated categories that were off the scale in the opposite direction, smaller not only than the genus and the species, but even than the shadowy variety. Everett Millais illustrated their relations as a set of embedded circles, with the "species" on the circumference, descending through "varieties," "strains," and "families," to the "individual" at the core.[136] Thus, while mirroring the hierarchy of scientific classification, these agriculturalists had also inverted it, retaining the formal authority of its structure but dispensing with that of its content. And if the titles of these transmogrified categories nevertheless echoed the language of natural history, the names of the groups of animals they classified proclaimed that they had been transplanted to a very different, more homely and practical sphere. Naturalists might refer to the Bovidae, or they might, if their intended audience was relatively broad or miscellaneous, refer to the ox tribe, but they never referred to families or tribes with names like Booth, Bates, or Cruikshank, let alone Emily, Wild Eyes, Lovely, or Cold Cream.

By means of such provocative adaptations, stockbreeders demonstrated that they were strong readers of latinate nomenclature: not only did they reinterpret its message to suit their own purposes, they also confidently emphasized that they had done so. After all, the goal of their enterprise was very different from that of natural history. The formal differentiation embodied in the hierarchy of breeds and sub-breeds was important to agriculturalists as a sign of their ability to manipulate nature directly, by reshaping the bodies of animals, rather than metaphorically through understanding and analysis. The improved breeds of the eighteenth and nineteenth centuries constituted a significant technological accomplishment; and so they were represented by their creators and boosters. For this reason, the relationship posited by stockbreeders between their breeds, tribes, and families and the genera, species, and varieties of natural history had to remain on the level of analogy.

As Darwin pointed out in *The Variation of Animals and Plants under Domestication,* breeders were as likely as naturalists to reject the notion that "natural species" and "domestic races" were identical, although for different reasons.[137] While naturalists emphasized the greater differentiation of wild species, breeders disparaged any suggestion that their strains were natural, as opposed to artificially fabricated. Even though the term *improvement* carried the connotation of inherent qualities, gradually teased out, eighteenth- and nineteenth-century stockbreeders understandably tended to stress the disjunction of their highly finished products with the raw zoological materials that nature had provided. A contrary emphasis on continuity would have inevitably undermined their claims about the value—and perhaps the very existence—of what they had to sell. Thus breeders walked a kind of figurative tightrope, implying that artificial breeds resembled natural species or varieties in their distinctiveness or individuation but that, at the same time, their development represented a violation of natural processes rather than an extension of them.

Despite their apparent high-handedness, these taxonomic transformations were, at least in intention, as respectful of zoological expertise as the more literalistic appropriation of latinate nomenclature. In a sense, the very boldness of the changes rung by stockbreeders on the practices of naturalists may have signaled a more thoroughgoing acknowledgment of scientific authority. By erecting a more precarious structure of assertions upon the foundation provided by zoological classification, the breeders implied a proportionately higher estimate of the solidity of that base. Technical nomenclature, and the structure that it expressed, were thus assumed as givens, part of the realm of established fact—secure enough to admit even the license of humor. When aficionados of domesticated animals indulged in binomial witticism, their targets were not ordinarily the labels themselves but the creatures that might prove unworthy of their ascribed dignity. Thus "*semi-homo canis*" satirized pets "who live with us nearly upon the footing of our fellow-man," while "Bos Composita" gently ridiculed the genealogical pretensions of shorthorn cattle.[138]

By assuming such stability, however, breeders also called it into question; the accolade contained in their imitation might be too high. Their deference to zoological authority was not itself entirely neutral, and the flow of influence it represented was therefore not entirely unidirectional. Neither the spirited transformations of cattle breeders nor the more slavish borrowings of dog fanciers precisely reproduced the nomenclature or the classificatory system of naturalists; in general, sound was preferred to sense. And this imperfect echo was then further distorted in its dissemination to a public that was very large and significantly different from the audience for technical natural history. If one consequence of this transfer—this analogy between kinds created by nature and kinds created by human effort—was to enhance the prestige and plausibility of domestic breeds, another, reciprocal consequence was to undermine that of zoological nomenclature, by impugning its absoluteness. Its adaptation to such a different purpose constituted a significant modification, perhaps the earnest of others still to come. And the willfully eccentric version of the language of classification offered by animal breeders constituted a commentary or interpretation as well as a reflection—a parody, unintended but nonetheless telling, as well as a transformation.

The undermining effect of these admiring appropriations may have been exacerbated because the level of taxonomic discrimination most relevant to animal breeding was the level most troublesome to zoology. Long before they helped lead Darwin to the formulation of his theory of evolution through natural selection, questions about how to define a species, whether the species or the genus was the fundamental unit of classification, whether varieties existed in the same sense that species did, and so forth, had bedeviled scientists. Despite the formal assurance projected by the vocabulary of classification, the species itself was a very uncertain category. Although comparing breeds to species might help to validate breeds in the eyes of fanciers unfamiliar with scientific controversy, the same juxtaposition further problematized the species in the eyes of naturalists. And, paradoxically, if breeders explored the causes of this instability, they might find their confidence in the authority of binomial nomenclature undermined, at the same time that their comparison of breed to species was better justified. The issues of purity and descent that troubled breeders were also problematical in nature. Crosses and hybrids, therefore, were as much the preoccupation of the museum and the natural history society as of the stockyard and the show floor.

3

Barring the Cross

As Charles Darwin pointed out in *On the Origin of Species,* the kind of discrete species that ambitious breeders of domesticated animals used as a touchstone was inconsistent with his theory of descent with modification—or, indeed, with any evolutionary theory, whether its mechanism was natural selection or not.[1] If species changed slowly but markedly through time—if no precise boundaries separated living forms from their extinct forebears—then the lines between similar living species correspondingly paled in significance. He summarized this argument in the final chapter of the *Origin:* "On the view that species are only strongly marked and permanent varieties, and that each species first existed as a variety, we can see why . . . no line of demarcation can be drawn between species, commonly supposed to have been produced by special acts of creation, and varieties which are acknowledged to have been produced by secondary laws."[2] With characteristic tact, Darwin offered this insight to the interested public as gently as possible. For example, he reassured traditional taxonomists, who might have feared that the very foundations of their lifework had been peremptorily dissolved, that not only would they be able to pursue their researches without disruption, but they would be able to pursue them with greater peace of mind and hence, implicitly, greater vigor. Once they accepted his theory, "they will not be incessantly haunted by the shadowy doubt whether this or that form be in essence a species. This, I feel sure, and I speak from experience, will be no slight relief."[3]

His assertion was welcomed by colleagues apparently only too happy to dispense with the species, at least in its most rigid and uncomfortable reification, which could, as Charles Lyell pointed out, cause embarrassment as well as intellectual confusion. He referred particularly to "the difficulty of defining . . . the terms 'species' and 'race', . . . [and to] the surprise of the unlearned . . . when they discover how wide is the difference of opinion" among experts.[4] Soon after the publication of the *Origin*, George Henry Lewes, explaining why "the zoologist sometimes . . . will class two animals as of different species, when they only differ in colour . . . [while] at other times he will class animals as belonging to the same species, although they differ in size, colour, shape, instincts, [and] habits," triumphantly revealed that "the reason is that the *thing* species does not exist."[5] Several decades later, an authoritative discussion of equine classification concluded that since "the old doctrine of the immutability of species and their separate and distinct creation is . . . not now held by the majority of modern naturalists, disputing . . . whether two closely allied animals are specifically or sub-specifically distinct is almost a waste of words."[6] The debunking of previous orthodoxy was fit even for the eyes and ears of children, as F. G. Aflalo demonstrated in one volume of a series called "The Library for Young Naturalists." Although a book devoted to "types" of animals was bound to invoke the notion of species with some frequency, Aflalo forewarned his juvenile readers that "It is . . . for reasons of convenience that men have invented species. Nature knows no such distinction."[7]

The energy of these rejections, as well as their reiteration over a period of half a century, suggests that their target nevertheless retained some of its prestige and power, even at the end of the Victorian period. And the conviction that species were somehow real—that in labeling a group of organisms with a latinate binomial, taxonomists were identifying an entity that had an existence independent of that naming process—flourished in spite of a striking absence of consensus about the nature of the entity in question.[8] It did not, after all, take the heresy of evolution to raise awkward questions about the species category. Some of these issues had emerged earlier in the rarefied reaches of philosophy or theory. For example, impermeably bounded species could seem inconsistent with one of the cherished dogmas of Enlightenment natural history—that is, the subtly graduated chain of being.[9] But the problems that most bedeviled naturalists arose on a very practical level: the assumption that species names represented essentially real and unchanging entities did not greatly assist in identifying them on the wing or on the hoof. Although, as Darwin later conceded, many species were "tolerably well-defined objects"—that is, like humans or hippopotami, they were relatively easy to distinguish from all other creatures— many species demonstrably were not.[10] On the contrary, like mice, horses, and weasels, they bore an incontestable and potentially confusing resemblance to numerous other animals.

The confidence of even the most orthodox naturalists in the essential reality of species was constantly challenged by the profusion and variety of the world. As a result, with their neat species-by-species format, eighteenth-century natural history manuals and encyclopedias contained many entries that ironically foregrounded the agonized doubts of their authors about where to draw the line. Of a polar bear displayed in London in the 1730s, Thomas Boreman speculated that since it differed from other bears in head shape as well as in size and color, "perhaps they are of a distinct Species."[11] The proliferation of antelopes stymied the anonymous author of *The Naturalist's Pocket Magazine,* who complained that while some authorities recognized "more than forty species, Buffon only enumerates thirteen, and seems inclined rather to consider these as varieties than as absolutely distinct species."[12] Still more confusingly, wrote David Low of the same animals, "at one point they are connected with the massive forms of the Bovine group, at another . . . with the Sheep, at another they pass into the Goats so nearly, that the line which separates the species scarcely forms a natural boundary."[13] Further decades of effort on the part of antelope taxonomists failed to improve the situation. In 1875 an audience at the London zoo heard that although "many attempts have been made to classify these animals," none had proved "satisfactory."[14]

George Shaw's zoological compendium was full of such puzzles. He worried that the Bactrian camel "so much resembles the Arabian that it might rather seem a permanent variety . . . than a distinct species"; that "the Glama [llama], the Paco, and the Vicuna, have sometimes been considered as the same species"; and that "the difference between the Rabbit and the Hare, though known from daily habit and inspection, is yet by no means easily described in words."[15] And if it was a matter of conviction that species themselves never changed, the frontiers ascribed to them seemed constantly on the move. Shaw noted that in the few decades since the publication of the twelfth edition of Linnaeus's *Systema Naturae,* the number of species within the genus *Sciurus* (squirrels) had increased from eleven to "near thirty."[16] He optimistically attributed this expansion to "the spirit of research among modern naturalists," but Stamford Raffles, writing after still further expansion of the sciurian ranks, more soberly warned of "the necessity of caution in multiplying the number of species in this genus on mere diversities of colour, as intermediate varieties will often be found to connect species apparently remote."[17]

The absolute species was equally elusive, in theory or in practice, whether it was the quarry of the most serious naturalists or the most casual popularizers. Authoritative consensus on particular hard cases was rare. Expert scrutiny tended rather to compound difficulties than to resolve them, since both the troublesome categories and the unmanageable information were artifacts of science rather than of nature. The more that was known about a given group of animals, the more challenging it became to separate them into plausible species.

The painstaking work of description and cataloguing that occupied most naturalists of the eighteenth and nineteenth centuries thus paradoxically turned out to impede the recognition of well-defined species, not to expedite it. The greater the number of traits attributed to a given group of animals, the more of those traits it was likely to share with similar groups, and the more difficult it became to decide which of them were truly significant for classification. But if they were forced to acknowledge the inadequacy of human intelligence and ingenuity to the tasks of defining the species category in the abstract and of recognizing its boundaries in the concrete, eighteenth- and nineteenth-century naturalists did not for that reason give up the search for distinctions that faith and logic persuaded them must nevertheless exist. Instead they looked to nature—standing in place of a still higher authority—for firmer guidance. Where learned examination of creatures, whether superficial or anatomical, might submerge species boundaries in a welter of bewildering and irrelevant detail, they hoped that a less encumbered gaze might discern a much simpler key.

Natural Antagonism

Although human beings had a difficult time drawing the line between similar creatures, it was frequently asserted, their confusion was not shared by their fellow creatures, whose concerns transcended mere intellectual curiosity. In the most important transactions of their lives, those involving the selection of reproductive partners, animals could be relied on to identify members of their own species and to avoid members of other species. They obeyed a rule starkly formulated by Robert Knox as, "Nature produces no mules; no hybrids, neither in man nor animals."[18] Edward Griffith thus attributed his inability to determine whether an animal he called the painted hyena was a new species or merely a variety of the striped or spotted hyena to the absence of such evidence. He argued that "the procreative power of any two animals . . . determine[s] this point . . . [but no] such opportunity has been afforded in the present instance."[19] Not only could animals see through the deceptive resemblances that confounded human investigators, but they were believed to be particularly sensitive to the boundaries separating their own groups from those which seemed most closely akin. In making these fine discriminations, animals were forwarding a higher purpose. As Thomas Bewick put it, "Nature has providently stopped the . . . propagation of these heterogeneous productions, to preserve, uncontaminated, the form of each animal; without which regulation, the races would in a short time be mixed with each other, and every creature, losing its original perfection, would rapidly degenerate."[20]

Works of natural history offered voluminous testimony to the desire of animals to avoid miscegenation, often citing a mutual repulsion between apparently similar creatures as persuasive evidence of specific difference. According to the zoologist John Fleming, "in a natural state, the *selective attribute* of the procreative instinct unerringly guides the individuals of a species towards each other, and a *preventive aversion* turns them with disgust from those of another kind."[21] Such reasoning allowed the confident recognition of a species boundary between domestic cattle and the buffalo of the Old World because "they will not copulate together, neither will the female buffaloes suffer a common calf to suck them; nor will the domestic cow permit the same from a young buffalo."[22]

The closer the apparent resemblance, often the greater the ascribed aversion, as though nature needed to deploy more stringent barriers in particularly ambiguous or liminal cases. Thus, "the wolf both externally and internally so nearly resembles the dog, that he seems modelled upon the same plan; and yet . . . so unlike are they in disposition that no two animals can have a more perfect antipathy to each other."[23] Wolves were reported to "be particularly partial to dogs" not as potential partners but as potential dinners.[24] The Edinburgh Natural History Society was informed that one punctilious African wild boar not only "refused to copulate with a common sow, but presently tore it to pieces."[25] And naturalists frequently remarked that "notwithstanding the general resemblance between . . . [the rabbit] and the hare, their habits and propensities are very different, . . . and they . . . seem to have a natural aversion for each other," to the extent that "a rabbit will live upon more friendly terms with a cat than a hare."[26]

These demonstrations were spectacular and, in their way, conclusive. But aversion was only a first barrier, and, as experience readily demonstrated, not invariably reliable. If it was breached, whether by some natural freak or, more likely, as a result of human interference, there were further defenses to reinforce it. The conditions of artificial constraint and proximity under which captive and domesticated animals lived provided ready opportunities for matings between ostensibly unlike animals. Even if the animals were themselves loathe to take advantage of these artificially created opportunities, they might receive strong encouragement to do so from their owners. Not all animals faithfully resisted such forceful if unnatural suggestions. Sometimes, when they succumbed, the lack of consequences only confirmed the point made by their initial reluctance. In support of his contention that "it was beyond the reach of human ability to exceed the limits prescribed by nature, by uniting two distinct species of aboriginal animals, and thereby producing a factitious one, capable of re-production," a contributor to the *Hippiatrist and Veterinary Journal* reported on his repeated attempts to cross a buck rabbit and a doe hare. Although he was so far successful in overcoming their natural antagonism that "in one instance . . . the sexual intercourse actually took place . . . , there was no issue."[27]

Such unambiguous confirmation of the separation of similar species could not, however, invariably be counted on. One kind of hybrid, the mule, far from being impossible or unlikely, had been intentionally and reliably produced by crossing horses and asses or donkeys for thousands of years.[28] But those seeking corroboration in nature for the existence of essential species had to understand even such ordinary workaday creatures as exceptional. The union of horses and donkeys might be predictably fertile, but it was often claimed that the hybrid issue of such unions were forbidden by divine and natural law to produce further offspring by mating with each other. As Philip Gosse explained, species were "distinct forms which are believed to have proceeded direct from the creating Hand of GOD . . . we know of no fixed principle on which to found our decisions [about the difference or identity of species], except the great law of nature, by which specific individuality is preserved,—that the progeny of mixed species shall not be fertile *inter se*."[29] For this reason the zoological, veterinary, and agricultural literature dealing with equine mules contained endless reassurances to the effect that "male and female mules and hinnys are absolutely sterile."[30]

If the products of horse-donkey crosses could be confidently relied upon to be sterile (although not to be celibate—their readiness to copulate inspired repeated musings on the nonfunctional in nature), some other hybrids between recognized species proved unnervingly prolific. Faced with this apparent anomaly, naturalists committed to fertile reproduction as the determinant of species boundaries might repudiate previously accepted limits, no matter how hallowed by tradition, rather than abandon their theory. As George Garrard suggested, "there are not so many species as some have imagined; but all . . . are of one species, which propagate with each other."[31] In this spirit, when Thomas Eyton mentioned at a meeting of the British Association that the hybrid offspring of Chinese geese and common domestic geese had been fertile among themselves, he concluded that "the deduction to be drawn from what I have stated . . . is self-evident; namely, that the above-mentioned birds are one and the same species." More colorfully, John Jones denounced those who adduced the vigorous and prolific produce of cock pheasants and domestic hens to "prove the falsity of the received opinion that mules will not breed," asserting that "this case proves no more than that pheasants and dung hill fowls are of the same species, like the fox and the dog."[32] A later refinement of this general argument relied on a sliding scale of hybridity, such as that produced by the French scientist Paul Broca, which ranged from the "agenesic" ("mongrels of the first generation, entirely unfertile"), such as mules and hinnys, through the "dysgenesic" ("mongrels of the first generation, nearly altogether sterile") and the "paragenesic" ("mongrels of the first generation having a partial fecundity"), to the "eugenesic" ("mongrels of the first generation entirely fertile"), represented by the offspring of the above-mentioned poultry.[33]

Opposites Attract

The wish to understand full hybridization—that is, the ability to produce offspring whose own mixed strain would be indefinitely prolific—as a natural or divine guarantee that the two original parents belonged to the same species was not, however, universal. Some relatively disinterested naturalists, less committed to the concept of inflexibly defined and divided species, contented themselves with describing the phenomena of hybridity, rather than interpreting them. In general, their observations were consistent with the grander claims of their essentialist colleagues, but they were more willing to note apparent anomalies and less eager to explain them away. Lyell commented that "in regard to the mammifers and birds, . . . no sexual union will take place between races . . . remote . . . in their habits and organization; and it is only in species that are very nearly allied that such unions produce offspring. It may be laid down as a general rule, admitting of very few exceptions . . . , that the hybrid progeny is steril, and there seem to be no well-authenticated examples of the continuance of the mule race beyond one generation."[34]

Nor was such judicious detachment the only alternative perspective on hybridization. Many of those interested in the animal kingdom found apparent violation of the laws of nature attractive rather than dismaying or repellent. They energetically sought out exceptions at every level of the neatly layered formulation by which animals readily recognized conspecifics, preferred to mate only with them, and, if forced or induced to violate this preference, produced no offspring or, in the worst case, sterile offspring. When any creature conceived an affection for one of different species—and the more different the better—it was therefore always news. Using the language of human romance, naturalists reported that a female zebra had been deceived by "a common Jack-ass" painted with stripes and had as a result "admitted its embraces"; without benefit of any human hoax, a male "Mongooz" became "fond of she cats; and even satisfied his desires."[35]

Humans themselves could be the objects or the originators of passions that transcended or violated the species barrier, although accounts of this kind were carefully distanced by skepticism or censure. The most common of such breaches, those involving farmyard animals, were much more likely to figure in a legal context than in works devoted to natural history or animal husbandry.[36] More remarkable events figured prominently in the zoological literature. At the end of the eighteenth century Charles White reported that orangutans "have been known to carry off negro-boys, girls and even women . . . as objects of brutal passion"; more than sixty years later the Anthropological Society republished Johann Friedrich Blumenbach's summary of travelers' accounts that "lascivious male apes attack women" who "perish miserably in the brutal embraces of their ravishers."[37]

Most authors who reported human-beast encounters were circumspect about the possibility of progeny. Nevertheless, the idea clearly exerted a certain appeal, which, although generally unacknowledged, was strong enough, according to a mid-seventeenth-century report, to have tempted one "poor miserable fellow" into bestiality with a monkey, "not out of any evil intention . . . , but only to procreat a Monster, with which . . . he might win his bread."[38] In an intellectual climate where such an effort seemed potentially rewarding, it was not surprising that Edward Tyson assured readers of his anatomy of a chimpanzee, that "notwithstanding our *Pygmie* does so much resemble a *Man* . . . : yet by no means do I look upon it as the Product of a *mixt* generation."[39] At the end of the eighteenth century White recorded rumors "that women have had offspring from such connection" and proposed that "supposing it to be true, it would be an object of inquiry, whether such offspring would propagate, or prove to be mules."[40] More boldly, a Victorian impresario advertised the hairy Julia Pastrana as "a hybrid, wherein the nature of woman predominates over the ourang-outangs."[41]

The issue of other transspecific amours, even between primates, could be reported with frank appreciation. A dead baby monkey, half Cape baboon

Definitely not a hybrid.

and half pig-tailed macaque, claimed as "the first instance of a *hybrid monkey* on record," was displayed at a meeting of the Asiatic Society of Bengal in 1863.[42] In 1878 the secretary of the Zoological Society of London proudly announced that a male macaque was the father of the hybrid offspring recently borne by a female mandrill, and would have been father also of a half-mangabey infant except that the expectant mother had unfortunately fallen to her death before her delivery.[43] A decade and a half earlier, according to the superindendent, the same zoo had witnessed the birth of a litter of bear cubs whose mother was a European brown bear and whose father was an American black bear.[44] Although inter-ursine crosses had been previously unreported in zoos or elsewhere, these infants proved the harbingers of many others, involving the grizzly bear and the polar bear as well as more familiar and accessible species.[45]

Evidence of hybrids involving domesticated animals was much easier to come by but nonetheless appeared worthy of repetition and comment. Readers of John LeKeux's popular survey of natural history were informed that male goats "will readily propagate with the ewe" and, more optimistically, that these sheep-goat "intercourses happen very frequently, and are sometimes prolific"; a later author advised that although such mixtures were not ordinarily produced in Britain, they could be viewed without much trouble in Paris.[46] Actual hybrid offspring between dogs and various species of wolves or jackals aroused enough interest to be reported in the press or noted in private journals, sometimes with the suggestion that the crosses might themselves be converted to human service. For example, animals bred from the dog and the maned wolf of South America were reported to be "excellent . . . for the chase."[47] Any serious treatment of equids had to include accounts not only of workaday mules but of practically all imaginable combinations between horses, ponies, donkeys, asses, zebras, and quaggas.[48] These manufactured creatures presented still more attractive practical possibilities than did canine hybrids. When in 1902 some officers sent King Edward VII a zebra-pony hybrid from South Africa, they had the animal gelded "thinking Her Majesty might use him, and a stallion hybrid is always a terror."[49] At about the same time it was reported of British-bred zebra hybrids of all kinds that "those which have been broken [to the saddle] have turned out perfectly satisfactory"; the development of zebra-donkey hybrids (to be called "zebroids") for African transport was seriously recommended.[50] And, although they spurned buffalo, domestic cattle uncomplainingly interbred with American and European bison, gayals, zebus, and yaks.[51]

So great was the fascination exerted by hybrid creatures that many impossible mixes were reported as fact, or lingered over regretfully as persistent superstitions. Thus Renaissance reports of crosses between bears and dogs were repeated (if only to be dismissed) throughout the nineteenth century, and Aristotle's account of "a hybrid race between the dog and the tiger" was similarly elaborated, before being characterized as "inadmissible," in "The Naturalist's Library."[52] But skepticism was hardly the invariable rule in such matters. At one time or another, nineteenth-century British readers were assured of the existence of "a hybridous race . . . between the domestic cat and the pine marten," a pair of puppies "the produce of a lioness and a true English mastiff," a stuffed animal in the Keswick museum "said to be between a racoon and a sheep," a ram with antlers about which "the presumption is that it is a mule, got between two animals often found together in our parks," and "an animal between a stag and a mare."[53] People who were credulously inclined might even use such creatures as the basis for further speculation, as when Charles Gould, formerly the official geological surveyor of Tasmania, suggested that "a cross between some equine and cervine species might readily result in a unicorn offspring."[54]

More plausible crosses were reported with greater regularity, whether or not they were more likely actually to occur. For example, in 1831 the fellows of the Zoological Society were informed that a doe rabbit had given birth to a litter of which half had been sired by a fellow rabbit and half by a hare.[55] A lone rabbit-hare hybrid appeared in the *Veterinary Examiner* soon afterwards.[56] By mid-century, reports of the "leporides," a numerous family of rabbit-hare crosses bred by a Frenchman variously referred to as Roux or Rouy, began to surface in the British press, their reputation enhanced by their prominent role in Broca's theory of hybridization.[57]

Reports of hybrids between the dog and the fox were similarly frequent, although never accurate. In a questionnaire about animal breeding that Darwin circulated in 1839, he matter-of-factly asked, "if a fox and hound were crossed with pointer-bitches, what would the effect be both in the first litter and in the successive ones of half-bred animals?"[58] Other authorities offered more substantial proofs. In 1824 the *Annals of Sporting* published an engraving of a "Portrait of a Cross of the Dog and the Fox," a prick-eared, bushy-tailed creature that had been "produced from a tan terrier canine bitch, by a tame dog fox, in the possession of Lord Cranley, at his Lordship's seat, near Guildford." The editors asserted that the visual evidence and the circumstantial detail would settle a question that, they admitted, "has hitherto occasioned much doubt in the minds of naturalists."[59] A veteran sporting authority testified that another "vulpo-canine bitch" had produced puppies with a male terrier, although not yet with one of her own hybrid kind.[60] Visitors to the Ipswich Museum could admire a "singular Hybrid between the Dog and the Fox," which occupied the most prominent position among its stuffed canids.[61] And a late Victorian correspondent of the *Kennel Gazette* recollected that "a veritable cross between a dog and a fox was exhibited at a Kennel Club Show . . . not many years back . . . in the open rough sheep dog class." Subsequently the same animal had "actually won the first prize in the collie class" at a New York dog show, where standards were apparently more flexible, or at least different.[62]

Even the last barrier defending immutable and immiscible species—the claim that interspecific hybrids were themselves sterile, and in no case could produce a continuing mixed offspring—was vulnerable to this contrary flood of anecdote. Equine mules provided the thin edge of the wedge. Although their sterility was conventionally cited as dogma, female mules very occasionally gave evidence to the contrary, producing foals sired by either a stallion or a jackass.[63] Thomas Pennant recorded the birth of a foal to a female mule in the rustic neighborhood of Forfar around 1760, regretting that "as there is a superstition in Scotland about these productions, the foal was put to death."[64] A mid-nineteenth-century agricultural encyclopedia listed several offspring born to female mules and male donkeys or horses within the previous centuries, not only in rural Britain, but in Santo Domingo and Malta.[65] The appeal of such reports was demonstrated by the energetic attempts made to discredit them as, for example, "having been made under some misconception, probably owing to the pony or horse which bred . . . strongly resembling a mule," or "always occur[ring] in some out-of-the-way district."[66]

In the strictest sense, however, the defenders of naturally delimited species could feel safe with these very anomalous half-mule offspring. Although they violated an alleged law of nature, the infrequency with which they did so could make them exceptions that proved rather than overthrew the rule. Because their interest resided precisely in the unlikeliness of their existence, the implications of their transgressiveness were severely limited. Their male parent was always a horse or a donkey, so that they were never presented as the product of two hybrids, nor were they claimed to mate with each other to produce an indefinitely fertile race of mules. But other kinds of hybrids were less constrained in their reproductive capacities. Most frequently remarked on were those between the dog and its closest wild relatives. Not only was "there no longer any room to doubt, that the Wolf and the Dog will . . . produce an intermediate species, capable of subsequent propagation," but, in addition, "hybrids between . . . the Dog and the Jackal . . . have . . . been proved to be thus fertile."[67] Most promiscuously prolific were domestic cattle *(Bos taurus)* and their allies, which produced continuously fecund offspring across the boundaries that divided not only species but genera.[68] Thus in 1884 A. D. Bartlett, the superintendent of the gardens of the Zoological Society of London, "had the pleasure of calling [the] attention" of the Society's Fellows to "some remarkable Bovine animals," the offspring of three generations of crosses, whose forebears included the gayal *(Bibos frontalis* in the nomenclature of the day), the zebu *(Bos indicus),* and the American bison *(Bison americanus).* On this basis he characterized "the belief, so general, that all hybrids or mules are barren" as "simply a stupid and ignorant prejudice."[69]

Pushing the Limits

While the accumulation of well-attested examples of fertile interspecific hybrids undermined the essentialist position, the very characterization of such animals as crosses tended to reify the category of species by acknowledging that the parents represented two different entities. But such technical concerns, however significant, did not entirely explain either the amount or the intensity of interest in reproductive mixture. The definitional problems posed by hybrids, as well as their sexual genesis, resonated with other Victorian preoccupations, of more general concern than the species ques-

An alleged hybrid between a dog and a fox.

tion. The most broadly based appeal of hybrids depended on the fact of mixture, not the degree of difference.

The most striking hybrids of the nineteenth century enjoyed the sustained attention of naturalists, as well as a spectacular if brief sojourn in the public eye. Thomas Atkins, the proprietor first of a travelling menagerie and later of the Liverpool Zoological Gardens, had the good fortune to possess a male lion and a female tiger who became enamored of each other and produced six litters of hybrid cubs in the 1820s and 1830s.[70] In pursuit of the largest possible gate, he advertised them in the most sensational possible terms. The posters printed for a visit to Cambridge flaunted headlines like "Unparalleled Attraction! Prodigies of Nature!!" and "Wonderful Phenomenon In Nature!" On the reverse side of one advertisement, Atkins described his wild animals in jolting anapests. Here too the hybrid cubs took first billing:

> The *Lion,* great Monarch of Beasts HERE in state,
> Behold! to a *Tigress* allied as his mate!
> Cohabiting fondly, a union whence came
> Three Cubs, a strange race, LION-TIGERS by name;
> The *first of their species,* for never had man
> Before met their like, since creation began,
> And seen by the KING, he was pleased to declare
> The acme they were of all wond'rous and rare.[71]

Incontestable bovine hybrids.

If the menagerist chose to characterize the appeal of his star attractions in the lowest common denominators of exoticism, novelty, and royal approval, zoologists attributed weightier significance to their hybrid charm. The death rate for all big cats was high in early-nineteenth-century zoos, but the lion-tigers were particularly unfortunate. Although at least one lived long enough to give "the most evident symptoms of its native and indomitable ferocity," most of them did not survive the loss of their milk teeth.[72] This sad vulnerability, which was widely obvious since Atkins ballyhooed the birth of each litter, had one silver lining: a steady supply of small hybrid corpses for dissection and preservation. And the demand for them was intense. In 1833, for example, when a cub died in Cambridge, its remains were immediately secured by the university's Museum of Zoology, along with a four-page memorandum asserting that their analysis would "furnish an important epoch in the annals of its [natural history's] progress to perfection."

But the answers to the two questions that most pricked the memorialist's curiosity would hardly have lived up to that portentous billing. They had already been answered many times for many hybrids with no such weighty consequences. "Did it appear that any of the properties of the body were precisely similar to those of the lion kind, or of the tiger kind, without a mixture; or, as in the mulatto, was there a general blending of properties? . . . was there any manifest defect in the conformation of the sexual organs, and if so was this defect sufficient to induce the belief of its inefficiency in the function of reproduction?"[73] Nevertheless, the charisma of combination insured that serious zoological museums continued to prize sibling specimens throughout the nineteenth century. The Cambridge University Museum of Zoology acquired a second skeleton in 1864; a lion-tiger hybrid was listed in the 1898 catalogue of Walter Rothschild's extensive private collection; and a 1906 survey of the holdings of the British Museum (Natural History) named the lion-tiger hybrid acquired in 1827 (one of the first of Atkins's short-lived cubs) as one of the "more important items" in the series suggestively entitled "Domesticated Animals, Hybrids, and Abnormalities."[74]

Hybrids between zoo inmates also aroused diffusely respectful scientific interest, even though, as Bartlett observed, some "antipathy exists in opposition to keeping a (so-called) mongrel race of any animal."[75] His parenthetical opinion of this reluctance was apparently widely shared by those in positions of authority, since, if hybridization between wild species was not usually foregrounded as zoo policy, it did occur with some frequency, invariably as the result of deliberate quartermastering on the part of zoo management. Clearly, no hybrid offspring could be produced if potentially compatible males and females were not confined together. Whatever the avowed purpose of such cohabitation, the production of young with mixed parentage was characterized as "the best results."[76] In addition to the bears, monkeys, and cattle already mentioned, nineteenth-century British zookeepers facilitated the birth of hybrid cats, lemurs, genets, civets, equids (mostly between wild asses and zebras, rather than horses), antelopes, deer, peccaries, and wallabies, among others.[77]

When menagerists overstepped the bounds of serendipity to institute an active program of hybridization, they ordinarily did so on ostensibly utilitarian grounds. One way of appropriating the resources of nature was to incorporate them into the domestic breeding pool. Wealthy amateurs of zoology who were also interested in agricultural improvement might therefore welcome an infusion of what was synecdochized as "blood" from wild relatives to correct the defects or enhance the virtues of ordinary farmyard animals. In its early days, for example, the Zoological Society of London maintained a suburban farm for exotic deer, cattle, goats, and similar stock, in addition to the urban gardens in Regent's Park. Among its purposes, according to the 1829 report of the Society's Council, was "effecting improvements in the quality or properties of those [domestic animals] which are used for the table."[78] Zoo farm inhabitants included a zebu bull, whose "services . . . might . . . at reasonable charge, and with every prospect of mutual advantage, be made available to the public," as well as a Wallachian ram, to be crossed especially with ewes of the Dorset breed, and several moufflon to be crossed with sheep in general.[79] In a similar spirit, the scientific subcommittee of the Bristol zoo resolved that their "Brahmin Cow [be put] to an English Bull when expedient."[80] The thirteenth Earl of Derby, sometime president of the Zoological Society, pursued a parallel policy in managing his large private menagerie. His crosses of various Indian cattle with domestic shorthorns apparently commanded a ready market, and he also had some success with ovine hybrids, but two peccaries, "procured . . . for the purpose of trying some experimental crosses" with local swine, "have not been found to answer the purpose."[81]

From the perspective of practical advantage, the Earl's peccaries represented the cumulative results of these endeavors more accurately than did his cattle. Whether they were perpetrated by public institutions, private fanciers, or the high Victorian acclimatization societies that included "the Perfection, Propagation, and Hybridisation of Races already domesticated" among the projected benefits of naturalizing exotic animals, most such enterprises seemed inspired at least as much by idle speculation as by any sober pragmatism, whether commercial or zoological.[82] If the proof of the pudding was in the eating, little in the way of exotic joints and chops ever came to the table.[83] Surveying several decades of such attempts, a critic who professed general sympathy nevertheless commented that "except as the production of a mere matter of curiosity, we cannot imagine what useful end would be reached by obtaining such crosses."[84]

By the end of the nineteenth century, hybrids occupied the attention of zoologists concerned with heredity, as well as those concerned with classification and description. Published evidence on the subject included the results of systematic experimentation as well as of anecdotal observation. Yet in their design, their execution, and their reception by the public, the most elaborately constructed experiments were apt to reveal an imaginative—even a playful—element, continuing evidence of sheer pleasure in combination and recombination. For example, at the 1900 show of the Royal Agricultural Society of England (RASE), John Cossar Ewart presented a substantial display of what he referred to as "zebra hybrids etc."[85] His exhibit represented the experimental labors of half a decade, the results of which he had summarized a year earlier in *The Penycuik Experiments*. This volume had been received enthusiastically by reviewers. In both popular periodicals and those directed to aficionados of natural science, medicine, sport, or agriculture, writers praised Ewart's practical and theoretical contributions. The *Times* reviewer recommended it to the "attention of stock-breeders not only at home but in all progressive countries," and the *Morning Post* called it an "intellectual treat."[86] At the beginning of the generously illustrated *Guide*, on sale to show visitors for a shilling, Ewart himself emphasized that his purpose in implementing these elaborate experiments, and in displaying them at the RASE show, was sober and practical.

Yet both the content of Ewart's experiments and the composition of his display suggest a supplementary source of appeal. The exhibit featured fifteen live equines, of whom a few were fairly ordinary specimens: Valda, "a chestnut polo pony," included because she was nursing a hybrid foal, and Fatimah, an "Arab filly," included for purposes of comparison. But most of their companions were first-generation hybrids, the offspring of the stallion Matopo, a Burchell's zebra whom Ewart described as "wonderfully quiet and friendly . . . except when herding mares," and a variety of domestic mares, including, besides Valda, a "bay cart-mare," a "bay Irish thorough-bred pony," a "Skewbald Iceland pony," an "Isle of Rum pony," a "black Shetland pony," and a "bay Irish roadster." (The preponderance of ponies reflected the relative heights of zebras and horses.) The appeal of the main performers was buttressed by less dramatic backup animals—pigeons, rab-

Above: Matopo, the stud zebra.
Below: Mulatto, an Isle of Rum pony, and her half-zebra colt Romulus.

bits, and cats—of analogously mixed extraction, as well as by skins and photographs of animals unable to make the journey.[87] This exotic display, amid so many ordinary cattle, sheep, pigs, and horses, would have exerted much of the attraction of a traveling menagerie, even without the elaborately graduated scale of striping displayed by the zebra hybrids. In addition, according to a visitor who had observed them playing in the fields at Penycuik, outside of Edinburgh, the hybrids were "the most charming and compactly built little animals possible."[88]

Ewart's mixing and matching of the denizens of his Scottish stables had not been directed merely to the gratification of idle curiosity. The *Guide* argued that a comparison of the patterns of the hybrids with those of non-hybrid half-siblings, the products of subsequent matings between their dams and stallions of similar domesticated breeds, would definitively disprove the widespread belief in telegony, or, as it was less technically termed, the influence of the previous sire.[89] It also suggested, somewhat less persuasively, that analysis of the stripes of hybrids might indicate the evolutionary relationship between the various equine species; thus Sir John, the offspring of Matopo and Tundra, the Iceland pony, was "interesting, because in his colouration he is probably a wonderfully accurate reproduction of the primeval horse."[90] If the experimental design dictated by the first objective, the testing of a straightforward hypothesis, had been relatively systematic, that dictated by the second must have been rather speculative, even capricious. By the end of the nineteenth century, hybrids of zebras with horses, ponies, and mules had been produced in sufficient numbers for it to have become clear that coat pattern was a volatile characteristic, and therefore an unlikely indicator of precise evolutionary relationships. As Ewart himself admitted of hybrids between the Burchell's zebra and the ass, "these crosses are sometimes . . . faintly striped . . . ; at other times . . . profusely striped."[91] He nevertheless continued to shuffle the genetic deck at Penycuik, eager to see what turned up.

Hybrids could thus offer a license for speculation. Darwin frequently recurred to the subject as he developed his theory of evolution by natural selection, mainly in support of his contention that it was impossible to distinguish consistently between varieties and species.[92] Although he discounted hybridization as the origin of new species in nature, he argued that some domestic species had multiple wild ancestors, especially the dog, among whose forebears he counted "two good species of wolf . . . , two or three other doubtful species of wolves; . . . at least one or two South American canine species; . . . several races or species of the jackal; and perhaps . . . one or more extinct species."[93] Others pushed further in this direction, claiming, for example, that horses, cattle, cats, and pigs were similarly amalgamated from several distinct wild species.[94] After the Mongolian wild horse or Przewalski horse was discovered in 1881, it was widely declared to be a hybrid, perhaps "between the kiang [an Asiatic wild ass] and the horse."[95] Reflecting back on those debates, one commentator asserted that the novel equid was "so like a bad mule" that it "might well be termed the missing link between horse and ass," thus merging hybrids with a still more compulsive source of Victorian fascination.[96]

The Mongrel Horde

The contemplation of hybrids inspired horror as well as pleasure, in parts that varied with the mood and predilections of the observer. Thus hybrids between more or less distinct species, whether wild or domesticated, were sometimes described with breathless appreciation as "extraordinary" and "remarkable."[97] At other times they were more matter-of-factly characterized as a "mule race," a "heterogeneous production," or a "mixed breed."[98] Partisans at the negative extreme invoked the language of moral disapprobation, identifying hybrids as the source of "deterioration," "disturbing effects," and "a confused chaos of mongrelism."[99] And naturalists were not the only ones to take positions. The transposition of this discussion to the subspecific sphere of domestic animal breeds, where the potential crosses were all demonstrably similar and where infertility was not an issue, might have been supposed to moderate its tone. But if the intellectual stakes diminished with the zoological prestige of the categories, the financial ante rose. Decades of painstaking segregation and improvement, it was feared, could be undone with a single cross. Breeders stiffly defended the elaborately documented labor represented by pedigreed livestock.

Their defense may have been made more energetic by the latent realization, seldom directly acknowledged although often only thinly veiled, that the maintenance of high genealogical standards was an issue for the breeders as well as the bred. An anthropomorphic salmon in Charles Kingsley's *The Water Babies* summarized the problem when she referred to her trout relatives as "ugly . . . brown . . . spotted . . . small . . . and degraded," before acknowledging that "I have actually known one of them propose to a lady salmon."[100] In a society that valued boundaries of all kinds, and that devoted great energy to establishing and defending them, taxonomic border areas and the intermediate animals that inhabited them inevitably appeared as analogies or extensions of similarly ambiguous human categories. Even the temperate Darwin revealed a sense of the special moral significance of intraspecific crosses. In the middle of his argument that species and varieties were indistinguishable, he consistently denominated the offspring of parents of different species as "hybrids" while denigrating the offspring of parents of different varieties as "mongrels."[101] This choice of word was not merely descriptive; the disparaging connotation was much stronger than any compensatory denotative precision. As Darwin's contemporary W. C. Spooner argued in a slightly different context, "although the term *mongrel* is probably correct as referring to a mixed breed, yet, as it is generally used as a term of reproach, it should not be fairly applied to those recognised breeds which, however mixed or mongrel might have been their origin, have yet by vigilance and skill become . . . almost as marked and vigorous and distinctive as the Anglo-Saxon race itself, . . . whose mixed ancestry no one is anxious to deny."[102]

Those with greater investments in the elaborately celebrated but somewhat insubstantial technology of animal breeding expressed themselves more forcefully. The results of unions that compromised their rigorous standard of isolation by what could be called "the ill-judged and unscientific introduction of alien blood" were excoriated in no uncertain terms.[103] The author of a pig-raising manual published in 1885 castigated "mongrel specimens of the breed" as "gaunt and inelegant," while a contemporary expert on pedigreed dogs condemned the puppies that resulted "where the animal is left to stalk uncared for at a time when the greatest watchfulness is required" as "graceless and unshapely."[104] One agricultural expert described crossing as "a national evil and a sin against society," especially if the cross-bred animals were themselves used for breeding what he called "a generation of mongrels."[105] "Blood" that would have been perfectly acceptable, even admirable or desirable, in its proper place became reprehensible when transposed into a different genealogical context. A chronicler of the heavy breeds of horses lamented the fact that "the original [Shire horse] stock" had been "sullied . . . by the infusion of light [legged] blood."[106] A prominent breeder of shorthorn cattle considered what was euphemistically referred to as "alloy blood" in a rival strain as "a stain in the pedigree of shorthorns" and "a disgrace to the breed."[107]

This aggressive insistence on purity may have reflected some unacknowledged uneasiness about the claims symbolized by pedigree. Even the most elaborately documented breed was wreathed in indeterminacy at every stage of its descent. Its remote wild ancestors were as elusive as the parentage of the individual animals recorded in its early breed books. And between these two stages, little could be confidently asserted about the original subgroup of domesticated animals, the unsung and unsingable founders of the breed, who had somehow become detached from the rest of their kind. The occasional brave spirit nevertheless alleged, for example, that "we have seen the Short-horns from the ancient race existing in Northumbria anterior to its conquest by William of Normandy . . . steadily improving . . . through their own blood alone, uncontaminated by any foreign element," although even he felt obligated to hedge a little: "or if occasionally so, to such small degrees as to be unrecognized in the predominating merits of the original race." [108] But most breed advocates admitted that attempts to identify the forebears of contemporary improved varieties were "more amusing than profitable or instructive." They regretfully settled for much more modest formulations, such as that "whatever its origin, the Galloway has been for a long time . . . very distinct," that "the old English and Irish Longhorn is . . . of unknown origins," and that "the Devon is believed to be of very ancient origin." [109]

The general rule for breeders was that "on no account should a cross be permitted," and those who "tampered with" a pure strain were castigated as "reckless" or even "radical." [110] Ensuring that the animals kept up their end was, however, not always easy. They might not wish to copulate with the mate chosen for them, preferring instead to follow the dictates of their own hearts or minds. For several reasons, concern about this kind of insubordination focused on female animals. Female reproductive capacity was more limited and vulnerable than that of males, so any wastage was more of a loss. Further, their proprietors were stuck with the fruit of their indiscretions while, once sown, wild male oats were someone else's problem. And, at least among otherwise healthy animals, it was only females who expressed reluctance to copulate. Breeders consistently testified that males were only too willing, needing on the contrary to be curbed in their enthusiasm, since too-frequent sexual intercourse was feared to undermine the constitution of both sire and offspring. For example, one authority on horses advised against pasturing a stallion with a herd of mares because in this situation "in six weeks, [he] will do himself more damage than in several years by moderate exercise." [111] An expert on pedigreed dogs similarly warned that "If you possess a champion dog . . . do not be tempted to stud him too much, or you may kill the goose which lays the eggs of gold. One bitch a fortnight is about as much as any dog can do, to have good stock and retain his constitution." [112]

Breeders therefore concentrated their disciplinary energies on female animals, despite the fact that throughout the nineteenth century most of them believed that the male parent dominated in shaping offspring. [113] Many shared the stronger conviction that the female parent was "limited, in some measure, to perform the same office that the earth does for vegetables . . . nothing more than a receptacle, in which are deposited the seeds of generation."[114] One expert, faced with explaining why, in that case, it was not "an easy thing to produce at once very perfect animals, provided that males of the right form could be obtained," preferred not to posit the mother as a signficant source of variation. Instead he had recourse to the fact that "the offspring will, to a greater or lesser extent, partake of the form and structure of the grandparents." And even if such an absolute assertion of male dominance needed modification in view of the obvious tendency of young animals to resemble both their parents, the consensus of breeders still reserved the more vigorous genetic role for the stud. The imagery of activity and passivity remained useful in the modified case; it suggested, for example, that "the male gives the locomotive, and the female the vital organs."[115] As it became available, the jargon of science was easily appropriated to express the conventional understanding of animal reproduction. Everett Millais translated it into the new terminology as follows: "that the male . . . does influence the epiblastic and mesoblastic structures largely, and all out of proportion to the female is undoubted."[116]

Subsidiary as their contribution to the final product might be once they were appropriately inseminated, females could nevertheless absolutely obstruct the realization of their proprietors' grand genealogical designs. As in human families, it was up to them to maintain both their own purity and the purity of their lineage. In the view of the breeder—as, indeed, in that of the paterfamilias—to compromise the one was to compromise the other. The literature of animal husbandry brimmed with worry about the undesirable hybrids or mongrels that might be produced even by females subject to the most careful confinement and control. Such concerns did not compromise the postulated dominance of males. Their superior strength meant, for example, that feminine physical resistance could predictably be overcome, if not by the prospective father alone, then as a result of collaboration between the male animal and the human breeder. Sometimes it was sufficient to bleed a "rebellious" female, "for the purpose of enfeebling her . . . and thus forcing her to become fruitful"; and with the most refractory animals, stud and owner could resort to what was euphemized as a "forced service."[117] The persistently troublesome avenues of compromise and contamination were, on the contrary, those that confirmed the weakness conventionally attributed to females, by allowing them to come under the physical or mental influence of unsuitable rival males.

Such misdirected affiliations could have protracted, even ineradicable consequences. The concept of telegony, which was almost universally believed by nineteenth-century breeders and fanciers and widely accepted within the zoological community, attributed to the "previous sire"—usually understood as the father of a female's first child—the power of influencing her subsequent offspring. The first serious public discussion of this subject occurred in 1820, when the original owner of the animal thereafter eponymously designated as "Lord Morton's mare" wrote the President of the Royal Society about her remarkable reproductive career. She was a chestnut of seven-eighths Arabian blood, whose first foal had been sired by a quagga that Lord Morton was attempting to domesticate. The similarity of this hybrid foal to both parents would not have inspired communication with the officers of great learned societies, but the fact that the mare's next two foals, both sired by a "very fine black Arabian horse," also bore "a striking resemblance to the quagga" was more surprising.[118]

Although the chestnut mare and her offspring were the most frequently adduced, perhaps because the best documented and most reliably attested, examples of this kind of time-lapse hybridization, they were far from the only ones to be publicized. A domestic sow first mated with a wild boar

The non-hybrid offspring of Lord Morton's mare still showed perplexing stripes.

continued to produce piglets "easily distinguished by their resemblance to the wild boar" in subsequent litters.[119] The grandchildren by other grand-sires of a short-haired domestic cat still sported the long silky fur of the Persian tom who had sired her first litter.[120] An Aberdeenshire heifer who had a calf by a Teeswater bull bore another "cross" to a bull of her own breed, and "popular belief" (reported, however, to an educated audience) suggested that "the children of [a widow's] *second* marriage may resemble their mother's *first* husband, both in bodily structure and mental power."[121] According to a well-regarded expert on pedigreed dogs, "the most careful and observant [breeders] have from their own experience recorded instances in proof of it, so that it is now an accepted fact."[122]

This persistent influence, whether merely "remarkable," as Darwin char-acterized it, or "very disturbing and confusing" in the words of a contem-porary surgeon, turned out to resist easy explanation.[123] The precise mechanism by which a female's first sexual partner left behind some con-tinually active physical trace of himself eluded investigators, although they suggested a variety of explanations. These competing hypotheses varied significantly in their details, but, as names like *inoculation, infection,* and *saturation* indicated, they concurred in emphasizing the appropriation, contamination, and even transformation of the virgin female by her original mate.[124] And, as animal breeders were quick to recognize, they all underlined the practical importance of guaranteeing female chastity and preventing "mésalliances."[125] Because "when a pure animal of any breed has once been pregnant to one of a different breed, she is herself a cross ever after," her subsequent offspring, even if appropriately sired, would inevitably "deterio-rate."[126] Indeed, in the most extreme cases of influence, they might not even have the chance to be born. According to a Polish traveler whose testimony was repeatedly cited by British authorities, "among the aborigines of certain countries, the effect of fruitful intercourse between a native female and an European male is to incapacitate the female from ever afterwards conceiving by a male of her own race."[127]

Occasionally, a free spirit tried to turn telegony to advantage. Thus a few adventurous breeders were reported to have put heifers of other strains to shorthorn bulls, in the hope that subsequent purebred offspring would retain some of "the excellence of the shorthorn," and to have put a Bedlington terrier bitch to a bulldog for a first mongrel litter, in the hope that subsequent purebred Bedlingtons would "have much stronger jaws than they would otherwise have had."[128] But most breeders strove to eliminate or contain the contamination that could result from miscegenation. The literature of animal husbandry was full of simple and straightforward prescriptions. When a bitch formed "an undesirable connection . . . breeders . . . at once put the strayed bitch down, or discard her from their kennels." [129] Breeders were advised "never . . . [to] commence a herd by breeding from a purchased cow," no matter how good she seemed on her own merits. She might have "been accidentally or otherwise served" by a cross-bred bull or a bull of a different pure breed, in which case "the risk is always there that she may throw calves with a cross-bred strain." [130] Still more fastidious was the suggestion that "the names of thoroughbred mares which have ever been covered by a half-bred horse, [or] by a Turk, Barb or Arabian . . . be rigidly excluded from 'The Stud Book.'" [131] So compelling were fears of covert hybridization that, although males were never seriously alleged to be subject to telegonous contamination, one "successful breeder" committed the "ultra-absurdity" of refusing to complete the purchase of a bull he admired after "he learnt that this bull had been running with cows of a different strain." [132]

Naturalists manifested a parallel fascination with telegony, and with the underlying issues of hybridization and purity. [133] It was considered a "subject of very considerable importance that has not received the scientific investigation that it merits," especially since it was "recognised by physiologists as affecting the human species." [134] This sustained attention was the more striking in that, however persuasive breeders and fanciers found the anecdotal literature, little of it met the standards of evidence to which zoologists ostensibly subscribed. The single example of Lord Morton's mare, celebrated as she had been, provided a rather shaky foundation for scientific theorizing, especially as there was a readily available alternative explanation for the appearance of her nonhybrid foals. Young horses and asses of many varieties show early, evanescent striping, and in some at least barring persists into adulthood, a fact that suggested to Darwin that the common ancestor of modern equids had been striped, although it did not convince him to abandon his belief in telegony. [135] Telegony continued to attract repeated attempts, often both expensive and cumbersome, at proof or disproof. As a reviewer of *The Penycuik Experiments* summarized the state of knowledge in 1899, "there is a considerable body of opinion, both practical and theoretical, for and against telegony." [136]

This same reviewer hoped that Ewart's ongoing contribution to the debate would be definitive. In designing his research, Ewart had gone back to the beginning. He tried to replicate the experience of Lord Morton's mare, choosing the zebra Matopo as the best available substitute for the original quagga, which had become extinct in the intervening decades. Using a variety of horse and pony mares, he bred each first to Matopo and then, for the second foal, to a stallion of her own or similar breed. In case after case, the result was the same: "neither Hector nor Argo afford any evidence in support of the widespread belief that a mare throws back to a previous sire"; "in no respect does Circus Girl support the belief in Telegony"; "in no single point has Kathleen ever suggested a zebra."[137] His results confirmed Ewart in the skepticism with which he had begun his work, and he referred to telegony and its adherents with a freedom bordering on contempt.[138]

Despite the strength of his negative evidence and his skeptical convictions, however, Ewart apparently felt, like many of his contemporaries, the attraction of a theory that emphasized the vulnerability of pure blood. At least, although others might claim that as a result of his work "all belief in the possibilities of its occurrence has been banished from the minds of scientists," Ewart could not bring himself absolutely to deny its possibility.[139] He therefore ended *The Penycuik Experiments* on an uncertain note—"it would be premature to come to any conclusion as to whether there is such a thing"—and suggested further trials with horse/zebra crosses as well as a long list of others, including a zebu bull with an Ayrshire cow, an Irish boar with a Tamworth sow, and a billy goat with a ewe.[140] Other zoologically inclined commentators proved similarly reluctant to dismiss telegony out of hand, even when they found nothing to say in its favor. For example, Millais, as part of his campaign to raise the scientific consciousness of his fellow breeders, characterized telegony as "almost . . . a myth."[141] Yet although "in a breeding experience of nearly 30 years" he had "never seen a case of telegony," and although he offered many alternative explanations for the instances commonly reported by his colleagues, he merely concluded that it was "exceedingly rare."[142]

Telegony may have been particularly difficult for dog breeders to dismiss because their bitches, like females of other species that give birth to multiple offspring, occasionally presented them with litters that actually did have two fathers, although each individual puppy only had one. Indeed, Millais suggested that "superfoetation," as this phenomenon was branded, explained many alleged instances of telegony. The resulting impure offspring could be attributed to a combination of human inattention and canine concupiscence. The mere possibility of such an occurrence would jeopardize the pedigree of the resulting litter, and so aspiring breeders were strongly advised that "for at least a week after the bitch has visited the dog, the precautions for isolating her must not be relaxed, or all her owner's hopes may be marred."[143] Such counsel mixed concern for breed purity with fear of female sexual appetite, which, if uncontrolled, was presumed always to

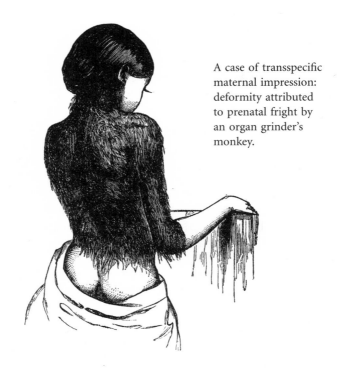

A case of transspecific maternal impression: deformity attributed to prenatal fright by an organ grinder's monkey.

lead to trouble. Breeders never imagined that bitches would choose dogs of their own breed, any more than they would terminate their heat early just because they had been satisfactorily impregnated. An early-nineteenth-century sportsman complained that "no convincing proof of satiety is ever displayed . . . and she presents herself equally to all," with the result that the largest of her suitors "is generally brought into action."[144] Similar suspicion of female sexual appetite and female indifference to breed boundaries underlay the repeated reports of superfoetation in humans, which invariably emphasized occasions when a woman had given birth to an apparently biracial set of twins, one black and one white.[145]

Even if they could be kept physically pure, however, females were still vulnerable to subornation and contamination. So powerful was the influence attributed to at least some males that it was believed that, as the result of a process called "imagination" or "mental impression," they might gain access to the reproductive organs through the eyes as well as in the ordinary way.[146] Like superfoetation, this phenomenon could be used to account for apparent cases of telegony. The so-called Leyswood Dingo, on display in 1897, was the daughter of pure dingo parents but nevertheless "decidedly not . . . typical" of the strain. Instead she resembled a wolf. One commentator attributed her eccentric looks to "mental impression," because her mother had had several previous litters by a wolf, who, in this interpretation, had left behind a strong impression not in her flesh but on her psyche.[147] Breeders anxious to protect their stock from mongrelization thus had to guard the minds as well as the bodies of their susceptible female animals.

Again the literature teemed with monitory examples. William Youatt, the most distinguished British veterinarian of the early Victorian period and a prolific writer on domestic animals, recounted the following story as an illustration of the need to control the imagination of "even so dull a beast as the cow": a certain cow "chanced to come in season, while pasturing on a field . . . out of which an ox jumped, and went with the cow, until she was brought home to the bull. The ox was white, with black spots, and horned. Mr. Mustard [the owner of the cow] had not a horned beast in his possession, nor one with any white on it. Nevertheless, the produce of the following spring was a black and white calf with horns."[148] The off-breed ox was not even completely male—that is, he had been castrated and therefore rendered incapable of procreation—but even so he apparently left his mark on the calf later conceived after intercourse with a properly pedigreed bull. And if females of relatively stolid species were so readily impressionable, it was not surprising that female dogs, considered particularly intelligent and excitable, caused still more trouble. Delabère Blaine, who was sometimes known as "the father of canine pathology," had a pug bitch whose constant companion was a white spaniel. All her litters were sired by pedigreed pugs, and all consisted of undeniably pug puppies—but one in each batch was white, a color that was rare and not desirable in that breed.[149]

Hybrid Vigor

If the ideology of purity was compelling in the abstract, however, it could prove problematic in real life. Principled abhorrence of mongrelization was nearly universal among those concerned with the business of animal husbandry, but their practical applications of that principle varied widely. Interpreted literally, it prescribed a standard of segregation that even the most fastidious breeders would have found difficult to satisfy. Indeed, too rigid a demand for purity would have threatened not only the reputation of well-regarded and long-established breeds but also the very reification expressed in the term *breed* itself. Although they might denounce contamination, taint, and bastardization with frequency and vigor, most realistic breeders knew that their chosen strains could not endure stringent scrutiny on these grounds.

The breeds with the most ancient pedigrees were among those most vulnerable to such deconstruction, so that their partisans often argued that past mongrelization did not preclude present purity. As a prolific authority on sporting dogs and horses asked in 1881, if such transformations were not possible, "where are we to find a certainly pure breed of dog, excepting perhaps the Bull-dog and the Mastiff?"[150] The frequently rehearsed mythology of the thoroughbred horse, which stressed the three foreign founding sires, made clear its origin in "an Eastern cross on native mares."[151] The fox terrier was equally "a manufacture—a compound of diverse elements," including the beagle and the bull terrier, as was the retriever, in the creation of which "Collie, Bull-dog, and even Hound blood was introduced."[152] The late Victorian chroniclers of Hereford cattle acknowledged that "like all our most valuable modern breeds, the Hereford would . . . appear . . . the result of various good sorts, both home-bred and foreign."[153] Indeed, the only reliable protection against disclosure of mixed ancestry was the absence of documentation. Thus the origins of the Yorkshire terrier were likely to remain "most obscure; for its originators—Yorkshire-like—were discreet enough to hold their own counsel."[154]

Occasionally, indeed, the ancestral mixture was not an ancient embarrassment, however submerged or justified by subsequent distinction, but a matter of continuing pride. The Colling brothers, early breeders of shorthorns, produced a strain founded on "a cross between the improved shorthorns and a polled Galloway cow," which was straightforwardly called "the *Alloy*, first . . . by way of contempt, afterward of commendation."[155] The first *Flock Book of Suffolk Sheep* noted happily if somewhat inconsistently that the early-nineteenth-century Southdown-Norfolk cross upon which the Suffolk was based "had been . . . perpetuated for generations with rigid adherence to purity of blood."[156] The first *Oxford Down Flock Book* announced the origin of the breed in "the deliberate crossing of two distinct types of sheep. It is perhaps not the only breed now in high favour which is founded on a cross, but it differs from almost all others inasmuch as this fact is its special pride and boast."[157]

Even crosses whose mongrel origins had not been transmuted into pedigreed gold had their uses. After all, equine mules were only the most striking of the many mongrels bred for routine use. Most of the horses that ubiquitously provided traction and transport were also crosses, and, as a late

Victorian agricultural textbook codified general agricultural practice, "*crosses between two distinct breeds* . . . make the best fatting animals."[158] Many highly satisfactory household pets similarly belonged to no particular breed, although less human discretion was involved in their design. But these uses did not include the production of further mixed generations. If pure descent had a practical value, in addition to its manifest rhetorical appeal, it was as a guarantee that the offspring of pedigreed animals would inherit the desirable characteristics of their parents.[159] As David Low commented, with reference to a cross between Zetland and Orkney sheep, "mixed races are rarely equal to those of pure descent . . . If we breed solely from [pure breeds] . . . we calculate securely on obtaining those varieties . . . ; but if we produce a mixed race, we can predicate nothing certainly regarding . . . the mongrel progeny."[160] Low's final claim was an understatement, since the concentrated reproductive force symbolized by breed status and pedigree was supposed to ensure that purebred sires and dams exercised a disproportionate influence over the resulting offspring, even when they were mated with more ordinary animals. So efficacious was social superiority, as embodied in pedigree, that it could tip the sexual scales that normally allotted the dominant role in shaping offspring to the male. Thus Youatt pointed out that the influence of "a highly bred cow will preponderate over that of the half-bred bull."[161]

This heightened power to shape progeny was called "prepotency." It was, of course, essentially comparative. That is, it offered a way to discriminate among breeds, as well as between pedigreed and nonpedigreed animals. It could therefore be used as a measure or confirmation of breed quality, especially since it could be tested in practice. The workings of prepotency seemed often simply to confirm the value of unsullied descent—to exemplify the rule by which "the most inbred parent generally influences the offspring to the greatest extent."[162] In some lineages it was alleged to operate so powerfully as completely to repel the taint of an injudicious cross. For example, the "old breed" of Suffolk horses exerted so strong an "influence" that "not one of the strains of alloy in the male line, could stand before [it]"; although foals showing the introduced mixture "in some cases stood for years, . . . sooner or later they died out."[163] A dealer in highly bred ponies boasted that one of his fillies was "so prepotent that, though she were sent to the best Clydesdale stallion in Scotland, she would throw a colt showing no cart-horse blood."[164] Similarly, according to Ewart, "the Jews, as a race, are more prepotent than the English—are better or purer bred."[165]

Other explanations were also adduced for this phenomenon, however, and they could complicate the relationship between prepotency and gene-

The pedigrees of the founding thoroughbred sires ran forward rather than backward.

alogically defined quality. Although prepotency usually developed as the result of inbreeding, it could also, according to one late-nineteenth-century authority, appear as an inexplicable "sport," especially in "a state of nature."[166] Indeed, domestication itself was generally considered to diminish prepotency. One admirer of the white Chillingham cattle dismissed the possibility of any lingering effects from former admixtures with neighborhood herds, "since the wild blood is proved to be so much more potent in its influence than the tame."[167] This effect could be somewhat paradoxically explained as a parallel result of alleged ancient descent. According to Millais, "the cause of the wild sire producing young more like himself than the domestic" was that domestic types were "of newer creation or evolution" and "consequently their type is less fixed." As examples he offered not only crosses of horses with zebras and quaggas, and of wolves with various breeds of dogs, but crosses between European people and members of darker human groups, which he considered to be both relatively old and relatively wild. If the father of a white woman's child was "a Mongol, a Polynesian, a Red Indian, or a Negro," he asserted, it "will resemble the sire to a much greater extent than where the white man is the father of the dark woman's child."[168] And many other possible reasons were adduced to explain the many apparent exceptions to the rule that connected prepotency and inbreeding. As G. H. Lewes summarized the confusion in 1856, "the causes . . . are various, some being connected with 'potency' or race, or individual superiority in age, vigour, &c; others being, in the present state of knowledge, not recognizable."[169]

The relationship between the prepotency of the individual and that of the breed or sub-breed to which he or she belonged was also problematic. Prepotency was inevitably demonstrated by individuals, because each offspring had only a single mother and a single father (except in cases of telegony and other unsanctioned interference). It was not always clear, however, in which direction that power flowed. In theory, if prepotency represented the culmination of long exclusive descent, individuals exercised it as representatives of their breed; but commentators often implied a reversed derivation. Prepotency varied greatly among individuals. According to one Victorian breed chronicler, "amongst animals similarly bred there are some bulls, and some cows too, that possess an immeasurably greater transmissible influence than . . . others."[170] Another suggested that such animals were the cause rather than the effect of particularly influential breeds—that "prepotency . . . exists not only in individual parents, but it establishes prepotent families."[171]

Shorthorn cattle provided similarly complicating testimony on this question. Beginning in the late eighteenth century, the shorthorn had enjoyed meteoric success; "other breeds melted away before, or became absorbed in it, leaving hardly a trace behind. The prepotency of shorthorn sires finds no parallel."[172] Although quintessentially domesticated, shorthorns of both sexes proved prepotent even over the ostensibly wild Chillingham cattle. An experiment, begun in 1875 and lasting over several decades, tried the effect of a Chillingham-shorthorn cross. The resulting offspring received short-horn names like Wild Rose and Wild Blossom, and by 1899 "most of the distinctive markings of the wild cattle have disappeared, while in physical configuration they have reverted to the short-horn type, until they are now becoming eligible for the Short-horn Herd Book."[173] Yet neither their tran-scendent quality nor their overwhelming power of influence could be at-tributed to long descent. Shorthorns were "a comparatively modern breed" and "of very mixed origin, possibly owing some of their merits to foreign blood."[174] Indeed, the mythology of the breed attributed its origin as well as its early success to the dominating influence of a single bull named Hubback, whose "prepotency [was] all the more remarkable when it is remembered that he was not in-bred in the least himself."[175] The relation between prepotency and exclusive descent was thus ultimately ambiguous, especially since, somewhat circularly, prepotency figured not only as a happy consequence of purity of breed but also as evidence that such purity in fact existed.

Nevertheless, the appeal of pure descent remained compelling in the face of this, as of other logical obstacles. Victorian breeders and fanciers ambi-tious to develop or consolidate a strain were routinely encouraged to follow the example of Robert Bakewell and other heroes of the eighteenth-century age of improvement. Without the guarantees that pedigree provided to their successors, these pioneering entrepreneurs had discovered that persistent inbreeding—referred to as "in and in"—provided the quickest method of fixing desirable characteristics and getting them to breed true. Exponents of this system were extremely fastidious in excluding alien blood that might contaminate their nascent strains. If blood relationship was necessary to ensure breed purity, then even a very slight degree of difference could produce mongrelization. That is, breed quality might be compromised by the blood of remote cousins, as well as by that of unrelated conspecifics. It was safer to stick to close kin. At the beginning, in-and-in advocates merely offered assurances "there can be *no danger* in breeding by the nearest affinities."[176] But their prescriptions gradually became more specific, sug-gesting that very radical measures might be required to eliminate the possibility of contamination: "mate sire to daughter and son to dam."[177] So satisfactory were the results of such close inbreeding in animals that they were repeatedly urged as justification for similarly hygienic practices among people—at least marriages between first cousins, if not between members of the same nuclear family—so long as "the parties" were not "both pre-disposed to the same disease."[178] The "extraordinary fear" with which people had traditionally regarded such unions was disparaged as the result of "ignorance" and "delusion."[179]

The more narrowly in-and-in advocates defined the universe of mates whose introduction within a strain would not vitiate its purity, however, the more resistance they provoked among fellow breeders. Some dissenters had always taken a high moral line; for example, the author of a mid-nineteenth-century agricultural manual used the language of faith and damnation to discourage inbreeding, warning that "it is an outrage upon nature which cannot fail . . . of producing baneful consequences."[180] But most critics relied on pragmatic rather than theological arguments, blaming inbreeding for a panoply of defects that afflicted elaborately pedigreed animals. As Millais put it, quoting a pseudonymous "medical man of scientific note" whose views were "accordingly . . . absolutely sound and correct," after protracted and exclusive inbreeding "the breed rapidly goes downhill, loses constitution . . . , becomes unable to withstand disease, and eventually becomes sterile."[181] The decline in fitness, as with inbred human families, was believed to involve mental as well as physical problems, including loss of "true courage and bottom" in hounds, diminution of "virility" in bulls and of "sexual appetite" in cows, as well as the complete disappearance of "sagacity" in dogs of all kinds.[182]

The invariable prescription for problems caused by inbreeding, whether dire or superficial, was "the introduction of fresh blood"—that is, a cross from outside the overbred line.[183] But in making this suggestion critics of in-and-in breeding did not mean to advocate mongrelization, whatever their opponents might think. On the contrary, they shared the appreciation for genealogical purity that had been intensifying throughout the nineteenth century, along with a consequent strong commitment to the defense of reasonably defined boundaries. The author of a late Victorian horse-breed-ing manual, who felt that "consanguineous intercourse has developed more imbecility in the human race than . . . any other cause," also asserted that "purity of blood is another essential, and perhaps the most important of all."[184] In The Book of the Pig, James Long similarly advocated scrupulous attention to lineage—even "the pedigree of an animal is not alone sufficient to distinguish it as being one of a great race"—while also asserting that "of all things in connection with stock breeding there is none . . . more necessary to avoid than . . . in-breeding."[185]

No matter how violently the parties to this debate disagreed, purity was always to be desired and mongrelization deplored. In a sense these opposed perspectives, and the divergent practices that embodied them, had become detached from the language in which they were expressed. Their superficial antagonism cloaked an underlying consensus. They did, however, differ on a subsidiary but far from trivial question. What inbreeders perceived as an issue of kind, crossers perceived as one of degree. Their argument was not about whether like and unlike animals should be mated but about the genealogical point at which likeness became unlikeness, or, to put it a different way, the point at which unlikeness stopped enhancing and began to threaten. At stake was not only where breed boundary lines should be drawn but also the rationale for drawing them, since, unlike that of the species, the definition of such artificial categories could be ascribed neither to nature nor to its author. The strength of the need to draw unambiguous boundaries between very similar creatures, combined with the transparent vulnerability of the resulting categories, produced heated rhetoric. And when the subjects under discussion were human—when the relationship between zoological and social categories appeared still closer—the temperature of discussion tended to rise again.

The Human Condition

Although the production of human children was arranged on grounds different from those that determined the pairing of pets and livestock, the two forms of breeding evinced an ineluctable relationship. Occasional bald statements to the effect that "man is a domestic animal" were supplemented by a constant stream of analogies, metaphors, and other figures of speech.[186] In people as in livestock, "blood" was the standard trope for a desirable and well-documented genetic package.[187] The genealogical guides to the gentry and aristocracy were familiarly referred to as "stud books." The biographer of one cattle breeder pushed the comparison further, introducing his subject in the standard herd-book form: "Thomas Bates, born Feb. 16, 1775; father, George Bates; mother, Diana, daughter of Thomas Moore; grandmother, Ann, daughter of Henry Blayney; . . . Such would be the entry . . . if the genealogies of the landed classes were registered on the same lines as those of pure bred shorthorns."[188] Often, indeed, animal and human pedigrees were presented together, as in the first Devon herd book, intended as a reference both to "the ages and pedigrees" of the cattle "together with the breeders and owners."[189] Millais warned inexperienced buyers to be wary of such juxtapositions, since "whenever I see a lot of the English nobility in a dog's pedigree I have . . . a shrewd feeling that the pedigree is better than the dog."[190] With middle-class owners the process could be reversed, allowing them to bask in the genealogical glow of their animal companions.[191] The same principles were routinely invoked to explain reproduction in both humans and domestic animals. Francis Galton coined the term *eugenics* because "we greatly want a brief word to express the science of improving stock . . . especially in the case of man."[192]

The similarity between human and animal pedigrees also figured in more general parallels between lineages, varieties, and even species. Eighteenth-century naturalists had concurred in dividing people into roughly geographical subgroups, although the number of subgroups, their precise territorial boundaries, and their distinguishing characteristics varied from authority to authority.[193] Linnaeus had recognized four main subdivisions—Americans, Europeans, Asiatics, and Africans—characterized by appearance, temperament, and habits. An early-nineteenth-century English rendering of the *Systema Naturae* thus described Americans as "Copper-coloured, choleric, erect. *Hair* black, straight, thick; *nostrils* wide, *face* harsh; *beard* scanty; *obstinate* . . . *Paints* himself . . . *Regulated* by customs," while Europeans were "Fair, sanguine, brawny. *Hair* yellow, brown, flowing; *eyes* blue; *gentle*, acute, inventive. *Covered* with close vestments. *Governed* by laws."[194] J. F. Blumenbach, who was regarded as a pioneering ancestor by many Victorian students of race, offered five categories, similarly based on geography and color. In his scheme Europeans were white, Mongolians yellow, Americans copper, Malays tawny, and Ethiopians black.[195] Buffon's elaborately subdivided survey of humanity confirmed the same general geographical pattern. According to a 1797 English version, among human variations, "the first and the most remarkable is the colour, the second is the form and size, and the third is the disposition"; the systematic alteration in these characteristics resulted from changing "climate, food, and manners."[196]

As with domesticated animals, such subdivisions, at whatever taxonomical level they were postulated, amounted to a reification, as well as a recognition or analysis, of difference. Subsequent classifications seldom diverged in their broad outlines, although they might employ alternative criteria. The facial angle promulgated as a classificatory device by the Dutch anatomist Peter Camper, which measured, roughly speaking, the slant from forehead to mouth, gave an average of 80 degress for Europeans, 75 degrees for Kalmucks (an Asiatic group), and 70 degrees for Africans.[197] More elaborate physiognomical analysis, such as that based on the work of Johann Kaspar Lavater, also replicated received geographical categories, with "the Frenchman, the Englishman, and the Circassian . . . formed in Nature's fairest shape; while the Calmouck and the Greenlander . . . [displayed] diminutive eyes, shapeless faces, and hollow nostrils."[198] Even the philological evidence of human difference adduced by some nineteenth-century ethnologists, although it was disparaged by the likes of Thomas Henry Huxley as insufficiently zoological, only confirmed the generally accepted arrangement.[199]

The lingering rhetoric of the chain of being combined with the emerging rhetoric of progress and evolution to ensure that classification was tantamount to ranking. Different taxonomists suggested different grounds for comparison. Some offered a standard based on physical prowess, claiming, for example, that "people in a savage state are less strong than those in a civilized."[200] Others valued the possession of a beard (a criterion that fortuitously accommodated sex and age, as well as race) or, like Galton, relied on measures of intelligence.[201] Racial groups were routinely distinguished according to their degree of civilization—Caucasians, in one account, having "an unbroken series of history, and . . . many varieties of civilization," while Mongols had "a less authentic history, and only one kind of limited civilization," and Africans "extremely little history and . . . seem to have retrograded in civilization."[202] According to the related criterion of theological development, Australians "imagine a being, spiteful, malevolent, but weak," and the "Negro's deity is more powerful, but not less hateful," while science had "purified the religion of Western Europe."[203]

The consequent hierarchy of race could be explained on a quasi-evolutionary basis, because, as one professor of medicine put it, "the European passes during uterine and infantile life, through stages . . . which are the adult characteristics of . . . the Mongolian and African."[204] But however the rankings were explained, they were entirely predictable. Every variety of evidence produced the same results. There was some room for argument about who should bring up the rear; although Africans were most frequently assigned this position, it was possible to protest that "the New Hollander, the Calmuck, the native American, are not superior" to them.[205] Such unflattering fungibility at the bottom only emphasized the unapproachable superiority of the occupants of the top position. As Charles White enthusiastically put it at the end of the eighteenth century, "we come at last to the white European," possessor, among other virtues, of "that nobly arched head, containing such a quantity of brain . . . , the prominent nose, and round projecting chin; those long, flowing . . . ringlets; that majestic beard, those . . . coral lips . . . that noble gait . . . the blush that overspreads the soft features . . . [and] on the bosom of the European woman, two . . . plump and sunny white hemispheres, tipt with vermillion."[206]

Consensus about the subgroups into which humankind was divided and their relative merits gave way to disagreement about the significance of the fact of difference. The most influential eighteenth-century taxonomists staked out a minimalist position by uniting all human races or varieties within a single species. Linnaeus thus included his four major races within *Homo sapiens,* along with the "Wild Man," described as "Four-footed, mute, hairy," and Blumenbach more assertively entitled one section of his treatise on human classification, "Five Principal Varieties of Mankind, One Species." [207] When, in 1764, Camper faced an audience that questioned whether Africans were "derived from the same parent stock" as other people, and suggested that they should rather be classed with apes, he "repelled these insinuations . . . with a warm and just indignation."[208] Such authoritative assertions trickled down to the larger community of naturalists, as when, in 1787, John MacFadzean informed his fellow members of the Edinburgh Natural History Society that "all Men . . . agree . . . in the particulars of their Structure . . . all [are] endowed with a reasoning faculty." He recognized a "difference in point of Ingenuity betwixt the Natives of Europe and Africa," but he considered it of no greater taxonomical significance than "the difference of intellectual powers in two Individuals of the same Nation." [209] And in 1813 James Cowles Prichard, like Blumenbach an inspiration to later anthropologists, declared that "the whole genus of Man contains but one species," despite diversities of color, hair, body type, and so forth.[210]

Naturalists of this persuasion criticized the notion of multiple human species with a regularity that suggested the persistent appeal of this alternative point of view. Indeed, Prichard declared in the preface to his study of human racial types that those of his countrymen who had previously explored the subject "have for the most part maintained . . . that there exist in mankind several distinct species."[211] In 1822 John Fleming asserted that although the existence of "permanent varieties" or races had "given rise to the belief that there are several species of the genus Homo, this opinion . . . is now generally abandoned."[212] And in his 1881 presidential address to the anthropology section of the British Association, William Henry Flower reminded his colleagues that "the varieties of man . . . have never so far separated as to answer to the physiological definition of species."[213]

Not all resistance to the idea of human unity took the form of explicit argument or declaration. The tendency to emphasize the differences between people rather than their similarity could be expressed tacitly, by the use of nomenclature. In this way, William Clift, the early-nineteenth-century curator of the Hunterian collection at the Royal College of Surgeons, accorded races the status of species when he noted that a rival anatomical museum contained specimens of *Homo Europaeus Britannici, Homo Asiaticus Sinensis, Homo Australasiae,* and *Homo Afer Antiquitatus* (a mummy), among others.[214] Indeed, even naturalists committed to the idea of a single human species sometimes revealed a linguistic tendency to exaggerate human subdivisions, labeling races or varieties with arcane, technical-sounding neologisms that formalized and substantiated the categories they denominated. For example, R. W. Latham, a follower of Prichard, was charged with writing "like a Frenchman" with a "perpetual effort after scientific . . . form," because he divided people into Iapetidae (Europeans), Mongolidae (Asians and Americans), and Atlantidae (Levantines and Africans), borrowing the latinate ending that signified a family in ordinary zoological usage.[215] Flower followed Huxley in separating "the Caucasion or white division" into fair-haired, blue-eyed Xanthochroi and the darker Melanchroi, even though "the two groups . . . agree so closely in all other anatomical characters."[216]

If Victorian advocates of human unity thus inadvertently expressed a zeitgeist increasingly inclined to emphasize the differences between people, exponents of the contrary position were correspondingly explicit and aggressive in their claims. During the 1860s, the nascent British anthropological community was riven by a struggle between the "ethnologicals," generally evolutionists and monogenists (believers in the common descent of all human varieties), and the "anthropologicals," generally anti-Darwinians and polygenists (believers in the independent origin of human varieties).[217] As a rule, although not inevitably, as Darwin pointed out in *The Descent of Man,* polygenists split people into multiple species and monogenists lumped them into one. Darwin also pointed out that, from an evolutionary point of view, in which one form "graduat[ed] insensibly" into another and in which the species, like other taxa, was merely a construction for the convenience of naturalists, "it is almost a matter of indifference whether the so-called races of man are thus designated [as 'races'] or are ranked as species or subspecies."[218] But his detachment was not characteristic of most participants in this heated debate. As James Hunt made clear in the presidential address that inaugurated the Anthropological Society of London in 1863, the abstractions of taxonomy could both explain and justify the political relations between Europeans and other people. Billed as a consideration of "the station to be assigned to [the Negro] in the genus *Homo,*" it argued that "there is as good a reason for classifying the Negro as a distinct species from the European, as . . . for making the ass a distinct species from the zebra." After a series of disparaging characterizations, Hunt concluded that "the Negro race can only be humanised and civilised by Europeans."[219]

The taxonomic relationship between Europeans and other human populations was mapped concretely as well as in words. Exhibitions inevitably emphasized the gap between observer and observed. Museums frequently displayed the remains of non-European human beings in ways that underlined their difference from Europeans or that suggested their greater affinity with other animals. Thus in 1766 a travelling collection of "curiosities" grouped a "Negro Child" with a "Monstrous Cat with 8 legs," a "Chicken's Foot with 6 Toes," a sloth, and an armadillo; a century later the Cambridge University anatomical collection listed separate entries for the "Tegumentary System or Skin" of the "Human" and the "Negro."[220] Flower urged the British Museum to follow the example of continental institutions, which mounted displays of "man as a zoological subject," so that Britons would learn to distinguish between "the blacks of the West Coast of Africa, the Kaffirs of Natal, the Lascars of Bombay, the Hindoos of Calcutta, the aborigines of Australia and . . . the Maoris of New Zealand" rather than "indiscriminately" applying "the contemptuous epithet of 'nigger.'"[221]

Live displays made similar points. The "only two Esquimaux Indians ever brought to this Kingdom" were referred to, in phrasing ordinarily reserved for animals, as a matched pair, "male and female." The advertisement for their visit to Derby in 1822 also stressed that they ate their food raw and that their only marriage ceremony was a dance; it pointed out that they were of special "National interest" because a colonizing expedition to their native land was then underway.[222] A Laotian girl was exhibited in 1883 as "Darwin's missing link," not only because she was unusually hairy but because she had prehensile feet and could pout like a chimpanzee.[223] Following the example of the 1867 Exposition Universelle in Paris, which included *tableaux-vivants* of North African subject peoples, British world's fairs routinely exhibited the indigenous inhabitants of the empire.[224] The Colonial and Indian Exhibition of 1886 accommodated almost a hundred exotic representatives of the peoples of India, Burma, Senegal, Canada, Cyprus, South Africa, Malaya, and Hong Kong.[225]

Although living Europeans were not normally put on display unless they possessed remarkable talents or physical anomalies, some Victorian practitioners of human taxonomy began to focus on increasingly subtle differences between ever smaller groups of people. Whatever its objective justification, the Enlightenment perception of European racial unity accorded ill with the global competition and intensifying national feeling of the late nineteenth century. Robert Knox thus accused such forerunners as Blumenbach and Prichard of having "misled the public mind" in this respect; he dedicated his own survey, *The Races of Men*, "to show that the European races . . . differ . . . as widely as the Negro does from the Bushman; the Caffre from the Hottentot, . . . the Esquimaux from the Basque."[226] Among the groups he reified with racial status were Saxons, Scandinavians, Celts, Slavs, Goths, and Italians. Others also found Knox's subdivisions persuasive. For example, researchers eager to examine the less superficial determinants of these intra-European racial differences could examine anatomical collections which, like that of the University of Cambridge, labeled

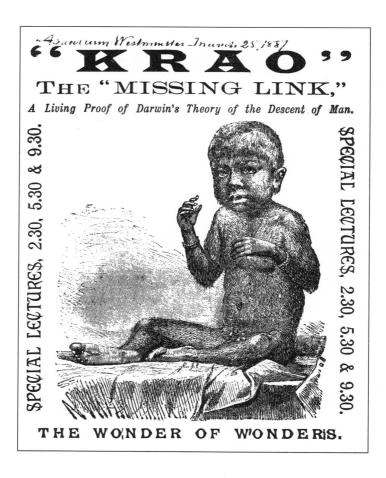

human skulls and skeletons according to their national origin, differentiating Swedish bones from German, and French from English.[227]

As with larger groups, such classifications inevitably involved ranking, and here too the ascribed hierarchy tended to reflect the political perspective of the observer. Knox attributed the perfection to which Georges Cuvier had brought the Musée d'Histoire Naturelle in Paris to the fact that, although a Frenchman, Cuvier was not a Celt "with whom order is the exception, disorder the rule"; instead he was "a Dane, or North German," with opposite mental characteristics.[228] Knox's appreciation of the "Saxon race," to which the English also belonged, was fulsome. It was "tall, powerful and athletic; . . . the strongest on earth; . . . the only absolutely fair [in color] race," and also first "in an abstract sense of justice, and a love of fair play." It was of special interest, he proclaimed, because it was "about to be the dominant race on earth." In his view, its only drawbacks were a certain physical disproportion, due to the massiveness of the Saxon torso, and a lack of artistic feeling and intellectual creativity.[229] Nor did Saxon boosters stop their comparisons at the national level, since Celts were as well represented in the British as in the French population. To this end they often replicated the established global racial hierarchy in insular terms. In his survey of *The Races of Britain,* John Beddoe speculated that "traces of some Mongoloid race" persisted in "Wales and the West of England" and that a certain prognathous "Gaelic type" had originated in Africa.[230] Even among Saxons invidious distinctions could be made. Galton claimed that the intelligence of the "Lowland Scotch and the English North-country men" was "decidedly a fraction of a grade superior to that of the ordinary English."[231]

Whenever human variation was reified as specific or subspecific difference, hybridization became an issue.[232] As with wild animals, the ability to produce indefinitely prolific mixed offspring was often perceived as an objective indicator of the "natural" distance between human groups. Consequently those commentators who posited the greatest differences between races were most likely to doubt or deny the existence of hybrid human strains. Such denials might extend even to mixtures of European peoples, although these alleged nonmergers were, prudently enough, more often attributed to psychological or political than to strictly biological causes. For example, Knox assured his readers that although the combination of "Teuton or Celt with a Scandinavian or Italian" was likely to be fruitful, nevertheless the "mingled races of Europe are not hybrids"—that in France, for example, "the Basque remains distinct from the population of old Gaul."[233] With regard to the major "racial" division of the British Isles, one correspondent of the *Lancet* asserted that "the Celts, rather than intermarry with

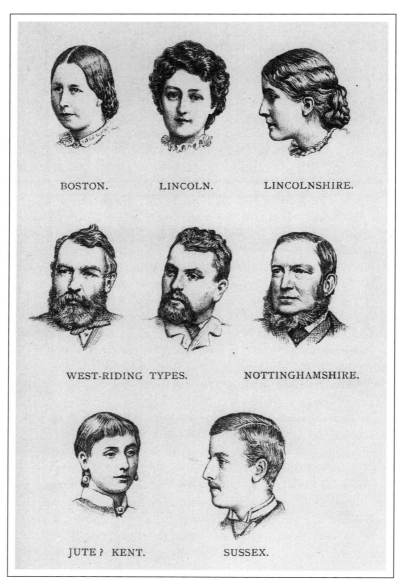

BOSTON.   LINCOLN.   LINCOLNSHIRE.

WEST-RIDING TYPES.   NOTTINGHAMSHIRE.

JUTE ? KENT.   SUSSEX.

Victorian racial discriminations became increasingly subtle.

the Saxon invaders, preferred to retire westward before them"; and a contributor to the *New Monthly Magazine* noted that in Ireland "there has been no amalgamation of the Celtic and Saxon races" because "they abhor each other cordially."[234]

Biological impossibility—that is, violation of the laws of nature—was more frequently invoked to gainsay the possibility of hybrid strains descended from the mixture of Europeans with non-Europeans. The testimony of Edward Long, whose *History of Jamaica* had been published in 1774, figured as powerful evidence well into the nineteenth century. Although it would have been difficult for the chronicler of a Caribbean island to deny the existence of mixed-race children, he made the more subtle claim that such offspring were themselves of diminished viability and fertility—that they were not fertile among themselves, and that children they produced with a member of one of their parent races were unlikely to live to a vigorous maturity.[235] Still less eugenesic, in Broca's terms, were crosses between Europeans and Australians, which, in spite of concrete evidence to the contrary, were alleged to be almost "perfectly sterile."[236] And despite the similarly concrete testimony provided by what Darwin called "the most complex crosses between Negroes, Indians, and Europeans" throughout South America and the Caribbean basin (crosses that provided the basis for standard reifications of terms like *mestizo, zambo,* and *quarteron* according to the amount of "blood" contributed by the various parent races), those committed to the distinctiveness and immiscibility of human races could proclaim, like Knox, that the admixture of alien blood was gradually disappearing from the mulatto population, which was "sure to . . . return to the aboriginal Indian population, from whom no good could come."[237]

If the possibility of human hybridization was thus variously debatable, there was more consensus about the unfortunate characteristics of any mixed populations that might result. Darwin asserted that although there were "many excellent and kind-hearted mulattos," South Americans "of complicated descent . . . seldom had . . . a good expression" and that "when two races, both low in the scale, are crossed, the progeny seems to be eminently bad." He viewed this phenomenon as an example of the reversion to the characteristics of wild ancestors often noted in domesticated animals.[238] In another echo of stockbreeding concerns, it was frequently suggested that although crosses with Europeans offered the most efficient, or perhaps "the only way to impart those elements of progress which are the glorious birthright of the highest races," the "intellectual and moral character of the European is deteriorated by the mixture of black or red blood" to the same degree.[239] Human hybrids were also considered liable to a range of physical and mental weaknesses; the only qualities occasionally adduced in their favor were good looks and the resistance to certain diseases. And even these were of unpredictable occurrence. Like mongrelization in animals, human mixing not only violated traditionally valued categories, but it introduced an uncontrollable and unsettling element of chance into established lineages. As the author of an anti-evolutionary broadside put it:

> And as the races intermix,
> You can't be certain about the chicks.[240]

The existence of hybrids or mongrels or crosses thus emphasized the existence of boundaries between groups and simultaneously obliterated them. The intensity of the aversion provoked by mixed creatures suggested the importance of the divisions thus called into question, whether they were zoological or political, agricultural or social. Often, indeed, that aversion could not be adequately expressed with the conventional rhetoric of purity and contamination; it required, in addition, the heightened register reserved for abominations. This language emphasized the extent to which hybrids violated the natural order, an order designed, it was variously alleged, precisely "to avoid filling the world with monsters"; to maintain "Distinction in her [nature's] Works" lest "a Line of Connection be drawn . . . uniting the Elephant and the Mouse"; or to prevent, as Lyell put it, "the successive *degeneracy,* rather than the perfectibility . . . , of certain classes of organic beings."[241] Yet as this rhetoric of repulsion emphasized the transgressive nature of hybrids, it also signaled their attractiveness. Border violations were appealing as well as threatening. After all, much more disconcerting monsters also played to large and appreciative audiences.

4

Out of Bounds

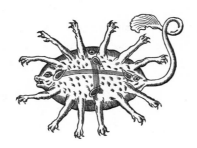

Oɴʟʏ ᴀ sᴍᴀʟʟ divergence from what seemed ordinary or natural sufficed to make a monster. G. H. Lewes offered a technical explanation of this transition away from normality in 1856, noting that when "children are . . . not only unlike their parents . . . [but] unlike their Species . . . we . . . call them Deformities or Monsters, because, while their Species is distinguished by having four legs, they themselves have six or more; while their Species possesses a complex brain, they are brainless, or have imperfect brains; while their Species is known by its cloven hoofs, they have solid hoofs, and so on."[1] Even a single exaggerated feature could put otherwise humdrum creatures over the line. Highly bred domestic animals, whether the prize livestock whose staggering obesity inspired the admiration of agricultural improvers, or the pet collies and bulldogs whose muzzles had been elongated or compressed for the delight of fanciers, were criticized as "not in fact perfect animals, but monsters, *i.e.*, deviations from, or modifications of, the natural type of the species."[2] William Henry Flower noted that "the power which . . . has led to the vast improvement seen in many domestic species over their wild progenitors has also ministered to . . . the perpetuation of monstrous forms"; he attributed this willful hereditary distortion to an innate human "propensity to *deform,*" which also caused people to dock the tails of horses, crop the ears of dogs, and encase their own feet in pointed shoes and their thoraxes in tight corsets.[3] The possessors of such self-inflicted mutilations might become monstrous themselves, like the one-legged Silas Wegg in Charles Dickens's *Our Mutual Friend.* The proprietor of a shop whose stock included "'Bones, warious. Skulls, warious. Preserved Indian baby. African ditto. Bottled preparations, warious,'" among many other stuffed and pickled items, reassured Wegg that, although his individual bones were not worth much, his skeleton "'might turn out valuable . . . as a Monstrosity.'"[4]

Unusual combinations, however produced, could similarly transfigure the ordinary. When late-eighteenth-century naturalists tentatively described the newly discovered platypus as an amalgam of bird, reptile, and mammal, they continued a Renaissance habit of interpreting American novelties as monstrous recombinations of familiar parts, analogous to the chimaeras and yales of medieval bestiaries. In this way, the opossum had been identified as half fox and half ape, the sloth as half bear and half ape, and the armadillo as a pig with a turtle's shell.[5] More dramatically, only a decade before the appearance in London of the first platypus remains, a broadside had featured a large illustration of "a Harpy . . . an Amphibious Monster now Alive in Spain," with the "face . . . of a Man, . . . two Horns like those of a Bull, Asses Ears, . . . a Main like to a Lions, . . . Breasts like a Woman, . . . Wings like a Bat, . . . horny Claws, . . . two Tails, one like a Serpents, the other terminates like a Dart."[6] Discoveries much less surprising than the platypus continued to seem similarly composite. In 1817, the first wapiti to be shown in Britain were described as resembling "the STAG, the HORSE, the OX, & the DROMEDARY."[7] In 1822, the "*lusus* of appalling aspect" produced by a Cumberland ewe was described in the *Annals of Sporting and Fancy Gazette* as an analogously unnatural pastiche of non-ovine kinds: "its ears resemble those of the fox . . . ; its lips are like the hare. Its eyeballs seem like duck eggs . . . It has teeth above and below, which is not the case with sheep."[8] (*Lusus naturae,* or sport of nature, was a standard term for monstrous births.)

The mere existence of hybrids whose parents were unlike in species, breed, or race testified to an analogous breaching of apparently natural boundaries, albeit by incontestably natural means. Such "spurious offspring" were stigmatized by both agriculturalists and naturalists, not only as mongrels but as "monsters";[9] even such unsurprising crosses as the mule could provoke allusions to the Old Testament, which "prohibited the coupling of different kinds of animals."[10] Less judgmental accounts of hybridization also associated it with monstrosity. Late-eighteenth-century anatomy lectures at the University of Edinburgh grouped "every mongrel" with animals "formed with one body to 2 heads [and] . . . 2 bodies to one head," and several decades later Busick Harwood similarly grouped mules and monsters in his Cambridge natural history syllabus.[11] One agricultural writer, more complexly conflating the reproductive behavior of people with that of their domesticated animals, interpreted the numerous hybrid breeds as proof that "the human race have fulfilled the threat of Caliban and 'peopled the isle with monsters.'"[12] In the same vein, Robert Knox declared that the "mulatto man or woman is a monstrosity of nature."[13]

The violation of biogeographical convention—by, for example, transporting a living or dead specimen from its original habitat to Europe—could also transform the unusual into the unnatural. And what became monstrous as a result of relocation in London could then, by a process of reverse association, seem monstrous even at home. Like the exotic animals whose remains were displayed in metropolitan museums, non-European human races were sometimes treated as monstrosities. Flower explained the fashionable deformities that made monsters of his stylish fellow citizens as "putting ourselves on a level . . . with . . . Australians, Botocudos, and Negroes."[14] A description, published in the *Lancet,* of a child stillborn with two partly attached bodies and one enormous head stressed that the mother, though a resident of Gloucestershire, had "remarkably dark and waved hair, with somewhat of the Ethiopic stamp about it."[15] At one further associative remove, the physical characteristics that distinguished human races figured as indications of monstrosity, rather than simple difference. In his discussion of the hereditary transmission of "mental defects . . . as well as corporeal," which Darwin consulted while formulating his theory of evolution by natural selection, Alexander Walker grouped the "distinctive character" of gypsies and the "Jewish countenance" with "supernumerary members on the hands and feet," the horny skin of a family of "porcupine men," and congenital idiocy.[16] A pocket manual published in 1874 by the British Association to guide amateur anthropologists in their investigations devoted an entire section to the "Abnormalities" or "*Natural* Deformities" presumed to characterize native peoples, including "*steatopyga,*" "peculiarity of the genital organs," "*Polydactylism,*" and "excessive hairiness."[17]

As many of these formulations indicate, monsters were understood, in the first instance, as exceptions to or violations of natural law. The deviations that characterized monsters, however, were both so various and, in some cases, so subtle as significantly to complicate this stock account. The causes to which they were attributed offered similarly inconsistent and destabilizing testimony. When an infant was born with physical equipment different from that of its parents, both the norm and the violation seemed clearly to fall within the realm of nature. But other monstrosities—not always easy to distinguish—had their genesis in human ingenuity, imagination, or violence. As a group, therefore, monsters were united not so much by physical deformity or eccentricity as by their common inability to fit or be fitted into the category of the ordinary—a category that was particularly liable to cultural and moral construction. Certainly individual monsters, especially if they were also human, often experienced the cultural consequences of their difference as painfully as the physical ones. Despite their natural (or unnatural) roots and associations, monstrosities functioned ultimately to define, and sometimes to redefine, social conventions and standards.

Monster Masses

The house of the monstrous contained many mansions. And from the seventeenth century onward, those mansions were ever more densely tenanted.[18] Monstrosities of all kinds poured into London, and many then circulated throughout the kingdom, stopping in towns and at country fairs. It is unlikely that more monsters were born or created at this time than had been the case in previous periods. The steady increase in the monster supply instead reflected an increase in demand. This demand was expressed by an audience difficult to reconstitute in detail; its variety was, however, suggested by the repeated references in advertisements to children, servants, and working people (who often benefited from concessionary admission fees) as well as to patrons of higher status and, presumably, greater sophistication. So great and reliable was the interest in the monstrous that, as the *Lancet* complained in 1825, the "rabid curiosity which afflicts the inhabitants of this metropolis" led "speculators . . . to turn this disease to account by the introduction of some extraordinary novelty," such as the "Living Skeleton" (a man afflicted with advanced pulmonary disease) then on display for "*half a crown* a dupe."[19] The fascination of the general public with monstrosities, as concretely demonstrated by its willingness to pay to view them, did not begin to wane until well into the Victorian period. Early in the twentieth century, Kenneth Grahame identified the "disappearance of freaks and monstrosities" as "perhaps the greatest change that has taken place in show-life in our generation."[20]

Exhibition occurred in several contexts. As long as they could command sufficient admirers, living monsters were displayed individually, often in the back room of a public house. For example, a "wonderful and strange English man, who is seven foot four inches and an half in hight" showed himself at the Sun in Cheapside in 1701, and "a surprising tall young woman, from . . . Surrey" showed herself at "a chandler's shop nr. Charing Cross" in 1753. In 1741 a female dwarf publicly advised noblemen that she was willing to visit their houses at a "price left to their own generosity."[21] When their novelty faded, or if their charms were relatively modest, monstrosities were exhibited in sideshows with others of their nondescript ilk. Thus "three legg'd cats" joined "a fire-eater, a giant, a dwarf, wilde beasts, [and] dancing dogs" at Midsummer Fair (in Cambridge) in 1711, and two monstrous rams, one with only one horn and one with six legs, graced the Bartholomew Fair midway in 1790.[22] By the nineteenth century, partially albinic Africans, known as "pied blacks," were among the most common human components of such assemblages.[23]

Preserved monstrosities were standard features of the first commercial museums, in which the exhibits functioned equally as agents of enlightenment and as public draws. According to the catalogue of a collection exhibited in the city of London during the 1760s, viewing "a monstrous Cat

Portrait of Mary Sabina, a "pied black."

with 8 legs" would help patrons "to facilitate the attainment of Natural Knowledge in one of its most considerable branches, and to gratify the imagination with one of its most elegant amusements."[24] An early-nine-teenth-century guide to the Leverian Museum, designed also as a natural history primer for children, stressed that an entire room was devoted to "*anatomical preparations* and monsters."[25] Since Benjamin Rackstrow's museum claimed an anatomical focus, albeit much blurred by the standard assortment of miscellaneous curiosities, monstrosities bulked dispropor-tionately large in its exhibits. An entire section was devoted to human "Miscarriages and Abortions," including "a Female Child . . . having no neck, the face looking directly upwards" and "another Monster . . . which had neither thighs nor legs, but the Body, as it were, lengthened, and tapering down to a point," while monsters "From Beasts," such as "A Two-headed Calf," "A Kitten with two lower jaws, and two tongues," a cyclopean puppy with a single eye in the middle of its forehead, and "A Pig, with a head resembling the Human," occupied their own space and cate-gory.[26]

Monsters were similarly integral to the collections of the nonprofit public museums that began to appear in the course of the eighteenth century. In its early years the British Museum included a department of "Modern Curiosities," which accommodated a lamb with two heads and a lizard with two tails; among the papers donated by its founder Hans Sloane were watercolors and drawings with such labels as "a dog with 5 leggs," "The Monstrous Head of an Abortive Calf resembling the Head of a Sheep," and "A Cock with . . . 2 asses [and] by both he dungeth."[27] Similar interests persisted (or recurred) through the Victorian period. Early-twentieth-cen-tury accounts of British Museum (Natural History) holdings noted that among the domesticated animals was a "collection of monstrosities . . . of quite recent origin" and that the equine series included several "horned horses," the significance of which was "at present inexplicable."[28] Curators of local museums echoed this predilection of their national model. At about the same period, for example, the Ipswich Museum displayed a two-headed lamb and a two-headed pig next to its monotremes and marsupials.[29]

More exclusive audiences found the same range of monstrosities com-pelling. By the early eighteenth century, anthologies of monsters designed for gentlemen's libraries had become a recognizable publishing genre.[30] Living monsters sometimes displayed themselves for the particular illumi-nation of the learned. Jacob Butler, an anatomy demonstrator at Cambridge and himself well over six feet tall, invited touring giants and dwarfs home to dine.[31] Preserved monsters occupied the center of attention at scientific gatherings. When Edward Tyson displayed a "Monstrous Cattling" with seven legs at a meeting of the Royal Society in 1702, his audience was so pleased that the specimen was ordered preserved in spirits of wine and added to the Society's repository; during the same period other investigators bombarded the Royal Society with accounts of giants, hermaphrodites, and acephalous infants, among other anomalies.[32]

Specialist private museums were similarly stocked. A manuscript catalogue of Joshua Brookes's anatomical collection thus began with "Monsters of Homo Europaeus Britannici," including infants born with too few and too many heads, a cast of a "foetus" extracted from the abdomen of a sixteen-year-old boy, and "many other species of human monstrosity, both uterine and extra-uterum."[33] When the Brookes museum was broken up in 1828, the Hunterian anatomical collection at the Royal College of Surgeons spent five shillings to acquire a cyclopean foetal lamb's head, although it was already richly provided with "the teratological series, or collection of congenital malformations of man and the lower animals," which, as Flower noted several generations later, "necessarily forms part of every general biological museum."[34] The annual reports on the condition of the anatomical museum at Cambridge singled out the acquisition of the "Skeleton of a monstrous Foal" in 1849 and of a "Double kitten" and the "Skeleton of a monstrous lamb" in 1864 as particularly notable; in 1895 the university zoological collection purchased a double-headed lamb from a local farmer.[35]

Expert Opinions

The naturalists and anatomists who guided and patronized such institutions tended to interpret monstrosities in the most restricted possible sense. In their view, monsters presented violations of natural law that were merely apparent, to be explained away by accurate analysis. Rather than indicating systematic chaos or caprice, therefore, anomalies offered opportunities to understand the natural order more fully and to define it with greater precision. The solution to the problem that they posed was to explain and classify them, to incorporate them into a newly elaborate and therefore newly powerful system. As a result of such efforts the light of common taxonomy—more often classification in terms of disease than of species—ultimately shone on some particular kinds of monsters. But monsters in the aggregate inevitably resisted definitive systematic integration, no matter how earnestly attempted. Capacious, motley, and irredeemably vernacular, the category "monster" proved invulnerable to expert analysis.

The basic principle that guided zoological investigators was straightforward. Richard Owen invoked his eighteenth-century predecessor John Hunter to the effect that "'*Monsters* are, like the Beings called Normal, subject to Constant rules'"; along the same lines, the romantic geologist Hugh Miller characterized "monstrosity" as a principle of deviation from the uniform skeletal architecture shared by "all higher non-degraded vertebrata."[36] After all, as the botanist Frederick Edward Hulme pointed out at the end of the nineteenth century, anomalous births could be seen as part of a large-scale, albeit undesigned, experiment; "amidst the millions of births in the animal creation there is scarcely any conceivable malformation, excess, or defect of parts, that has not at some time occurred."[37] Monstrosities offered, in the words of two contemporary teratologists, "forbidden sight of the secret work-room of nature."[38]

Although teratology seemed like a quintessentially modern endeavor, it was not without forerunners. Efforts to understand anomalies long predated the monster population explosion of the late seventeenth century. Along with bestiaries and cabinets stuffed with miscellaneous accumulations of natural and artificial curiosities, the monster literature of the Renaissance was routinely disparaged by enlightened later naturalists, on grounds of both credulity and lack of system.[39] These sweeping denunciations, like those directed at bestiaries and cabinets, were, however, somewhat partial and unfair. Such treatises as Aldrovandi's *Monstrorum Historia* offered more than jumbled illustrations of their sensational subjects. Aldrovandi explicitly suggested possible causes of monstrosity in an extended discussion of human generation and the development of the fetus, and the order in which he presented his monsters also constituted both an implicit etiology and an implicit taxonomy. He grouped a man with rabbit ears with a man with the tail and legs of a lion and several subtler examples of humans with transspecific characteristics, such as a hairy-faced girl and a man who walked on all fours; a fish with a human face with joined twins "*partus humanus, & caninus*"; and a two-headed pig with a variety of other bicephalous animals.[40] His French contemporary, the surgeon Ambroise Paré, similarly mixed the surprising and the impossible, including double monsters and such exotic animals as the ostrich and the elephant, along with incompletely human creatures like mermaids and satyrs. He distinguished, however, between monsters, which were merely "outside the course of nature," and marvels, which were "completely against nature."[41] Like Aldrovandi, he attributed such prodigies to a variety of causes, ranging from too much or too little seed in conception, through injuries to the mother during pregnancy, to the wrath of God.

These allegedly benighted efforts at explanation and analysis nevertheless evinced a certain staying power.[42] Through the nineteenth century, technical discussions of monstrosity continued to take account of the subterranean persistence of more eclectic earlier understandings, if only to dismiss them. In 1826 the Regius professor of medicine at Oxford found it necessary to assure his students that "an individual sheep may have two heads . . . but you will never find that one of those heads shall be the head of a sheep, the other the head of a horse," and, more generally, that "there is no ground . . . for supposing that nature has ever produced such an individual as a chimera or centaur."[43] A few years later, nevertheless, a contributor to the *Lancet* seemed to admit the possibility, by labeling his title anomaly "Hen with a Human Face." The transgressive affinity was more than skin deep, since she was also reported to diverge from other chickens in temperament, seeming "more fond of the society of men than of that of fowls."[44] A lecture delivered in 1840 to students at the Middlesex Hospital School of Medicine criticized not only the "gross superstition and credulity" of those who attributed monstrosities to divine or demonic intervention but also the lesser intellectual lapses of "the public, and also . . . the profession" who saw them as "a whimsical deviation of Nature from her accustomed laws." This peremptory dismissal prefaced a discussion of "the present comparative state of perfection" to which teratology had been carried by continental investigators.[45] Addressing a nonspecialist audience at the end of the nineteenth century, Hulme stressed that although the Bushmen were probably modern representatives of the "cave-dwelling, reptile-eating" pygmies de-

Hen with a human face.

scribed by Herodotus and Pliny, "such monsters as Aldrovandus figures are utterly impossible."[46]

One reason for the durability of these disparaged traditional views was the difficulty of developing persuasive replacements. At least in 1840, even teratology boosters could claim only that their field enjoyed a relative— rather than an absolute state—of perfection. For monstrosities to be definitively coopted into the service of science, it was necessary not only to describe them—a process in which eighteenth- and nineteenth-century naturalists merely extended and refined the labors of their allegedly unsystematic predecessors—but also to make sense of them; that is, to classify them and to explain why they occurred. Despite the apparently rapid advance of teratology—signaled by the shift in the primary reference of the word itself, from the seventeenth-century sense of a narrative dealing with prodigies and the marvelous, to the nineteenth-century sense of the scientific study of monstrosity—eighteenth- and nineteenth-century researchers could provide only tentative and inconsistent solutions for these problems.[47] Their efforts testified as much to their desire to incorporate monstrosity into the system of nature as to their understanding of anomalous creatures.

One reason for their difficulty was that taxonomy and causation turned out to be impossible to separate. Rather than representing a distinct intermediate stage between description and explanation, most systematic attempts to classify monstrosities depended heavily on accounts of their genesis. This chicken-and-egg intermingling was polemic as well as heuristic, since both classification and explanation exemplified self-consciously rational alternatives to traditional modes of understanding. The eighteenth-century German biologist Caspar Friedrich Wolff explained his division of anomalies into six categories—single monsters, dicephalous monsters, pyopagus monsters (two bodies joined back to back), craniopagus monsters (two bodies joined by the heads), double parasitic monsters, and tricephalous monsters—with reference to embryological development, as opposed to the "displeasure of the Creator."[48] John Hunter suggested that "of monsters there are two principle classes, viz. Duplicity of Parts and Deficiency of Parts; and there is a third class, viz. Bad Formation," each caused by a specific "defect in the first arrangement of the original matter"; these, according to Richard Owen's subsequent gloss, could be subdivided according to both the species and the body part primarily affected, each of which "has a disposition to deviate from Nature in a manner peculiar to itself."[49] Early in the nineteenth century Johann Friedrich Meckel insisted that monstrosities be classified not on the basis of their form but of "the type of deviation of the formative force." On this basis he proposed four major categories: monsters resulting from insufficient generative energy; monsters resulting from excessive generative energy; monsters resulting from aberration of form and of position; and double monsters.[50]

Teratologists continued to revise and extend their classificatory systems through the nineteenth century; in addition, following the example of Etienne Geoffroy Saint-Hilaire and his son Isidore, they attempted to supplement nature's raw material by fabricating their own monsters in the laboratory.[51] The results of all this activity and assertion were not, however, perceived as completely satisfactory. Sometimes this unease was expressed obliquely. Naturalists who desired to make sense of monstrosity might fall back on an alternative classification, simply ignoring the speculations of embryologists. Thus the dichotomy between wild animals and domesticated animals, which also strongly influenced conventional zoological taxonomy, was invoked to explain and delimit teratogenesis. It was frequently asserted that livestock and pet species were much more likely to produce monstrosities than were wild animals. There was less consensus about the reason for this disparity, although developmental explanations were generally rejected in favor of environmental ones. For example, one British interpreter of Buffon attributed anomalies to the sexual exhaustion of overused stud animals, whose "first productions . . . will be strong and vigorous" but from whose later "sluggish amours insipid beings must proceed," while another compared the "weak, degenerate race" and "great proportion of Monsters" characteristic of domesticated animals with the effects of civilization on "rude uncultivated savages."[52] John Fleming blamed the numerous monstrous births among domesticated animals on the "excess nourishment" that they were likely to receive.[53] In his early evolutionary speculations, Darwin pondered the possibility of "those parts being more easily mostrified, which last produced"; he thought this tendency would produce frequent insanity among humans "in civilized countries" and a general volatility of physical characteristics among animals subjected to intensive selection by breeders.[54] Medical practitioners with a more immediate interest in the subject might simply disregard the fine distinctions offered by teratologists, retreating instead to crude binary oppositions. For example, a surgeon writing in the *Lancet* in 1860, eager to persuade his colleagues to use prosthetics in treating viable monsters, asserted that "the great and fundamental classification of deformities is into the congenital and the non-congenital."[55]

Other critics of teratological systematics were more explicit, claiming that the Saint-Hilaires and their competitors had prematurely erected elaborate taxonomical structures before establishing a firm foundation of understanding. As one contributor to the *Lancet* put it in 1840, "notwithstanding system upon system has been piled, many much in accordance with and others diametrically opposed to each other, each and all professing to unravel the hidden mystery, they have been satisfactory alone to the minds of their founders."[56] A decade later the author of the article on teratology in an anatomical encyclopedia explained his decision to "avoid entering into a full critical examination of the systems propounded by Licetus, Huber, Wigtel, Malacarne, Buffon, Blumenbach, Breschet, Geoffroy St. Hilaire, Gurlt, Otto, and Bischoff" on similar grounds: "no suitable classification of monstrosities can be given, and the efforts to this end may be regarded as failures."[57] Looking back at the pioneers of teratology from the end of the nineteenth century, John Cleland, the professor of anatomy at the University of Glasgow, dismissed their classificatory efforts in a condescending comparison with the more powerful understanding of his own contemporaries. He asserted that, since "investigation has entered on a more strictly causal stage," the "thoughtful inquirer can be no longer content with merely cataloguing deviations and bestowing on them sesquipedalian names such as were perhaps justifiable in the days of the elder St. Hilaire," and, further, that "the advantage which biological doctrine has hitherto derived from teratology has not been great."[58]

Without benefit of hindsight, however, Cleland's predecessors had assumed the contrary. They might differ about the systematic significance of monstrosity but not about its existence as a significant category. Indeed, its very indeterminacy of meaning made it useful to the widest possible range of combatants. Monsters could be mustered in support of almost any point of view. As a result, they figured on both sides of some of the most protracted and hotly contested debates in nineteenth-century biology. Anatomists, for example, deployed monsters in their struggle to distinguish their dynamic discipline from the more traditional pursuits of zoological systematizers. As William Lawrence assured the Royal College of Surgeons in 1817, "the consideration of monstrous productions belongs to pathology, and physiology, rather than to natural history."[59]

Those committed to taxonomy were, however, undeterred by such territorial claims. They continued to cite the tendency of monstrous births to resemble perfect specimens of other species as a confirmation of the hierarchical and linear chain or scale of nature, long after that organizational paradigm had been supplanted by the more complex and at least relatively egalitarian tree or bush. That is, these transspecific resemblances were characterized in terms of ascent or descent, generally the latter: although "cases are much more rare in which inferior animals resemble the higher" (one clear example of this was the absence of the tail in some carnivores, such as Manx cats), most monstrosities "realise, in some of [their] organs, the condition that is met with among the inferior classes."[60] Since people occupied the top rung of all classificatory ladders, almost all human deformities could be explained in terms of taxonomic slumming, starting with the transparently named harelip. Babies born with hands and feet attached directly to the torso similarly represented a reversion to the cetacean condition, while those with a cloaca or single excretory orifice recalled birds and reptiles.[61] The most frequently adduced explanation of this patterned atavism or reversion—Meckel's ontogeny, according to which each developing fetus passed through a series of stages equivalent to the adult forms of lower classes or orders—also assumed that animal types could be arranged in a single graduated rank.[62]

In addition to explaining the development of nonstandard individuals, theories of monstrosity helped account for the emergence of new varieties and species. In general, it was difficult to draw the line between mild monstrosities and distinctive variations. As Darwin pointed out in his early notebooks, far from being inevitably sterile, the "most monstrous form has a tendency to propagate, as well as diseases"; indeed, not all monstrosities deserved to be viewed in the negative light shed by this association, since "even a deformity may be looked at as the best attempt of nature under certain very unfavourable conditions."[63] Some highly prized fancy breeds, therefore, could be viewed as having been established by the selection and propagation of the monstrous. Writing to Darwin, Edward Blyth proposed a taxonomical acknowledgment of this possibility: "the designation *breed* should be restricted to [strains descended from] . . . *normal varieties*," while strains with no "prototypes in wild nature," such as long-eared goats, angora cats, double-combed chickens, and poodles, should be grouped under the rubric of "*abnormal* varieties."[64]

If reasonable people could differ about the worth and origin of such breeds, they disagreed even more strongly about the origin (if not the worth) of wild species. Because of the analogy between the development of new breeds by people and the emergence of new species in nature, made powerfully explicit in the opening pages of *On the Origin of Species*, and also widely assumed in agricultural and zoological considerations of the subject, monstrosities played an equally significant role in this larger controversy.[65] Not all of this proto- or para-evolutionary discussion displayed the theoretical and chronological sweep of the debate that focused on Darwin's work. But at the least, the accumulating fossil record convinced many nineteenth-century naturalists that during previous epochs the earth had supported very different assemblages of plants and animals, and that some account was needed of the relation between those earlier floras and faunas and those of their own time.[66] Despite, or perhaps because of, the lack of consensus about their genesis and transmission, monstrosities figured as examples of the way that such alterations both might and might not have occurred. Thus for the conservative Richard Owen, rooting speciation in teratology offered a way to circumscribe the potential for organic change; no alteration, however abrupt or startling, could be really new.[67] Darwin—whose evolutionary theory on the contrary envisioned unlimited, snowballing transmutation, influenced as much by shifts in environmental conditions as by internal organismic constraints—dismissed monstrosities as trivial if sometimes striking departures from species norms. In *On the Origin of Species* he compared them to larval insects, which, although superficially very different from the adult forms, were nevertheless intrinsically the same.[68] In his subsequent works he further minimized the evolutionary implications of monstrosities, treating them rather as single characteristics, analogous to unusual coloration, than as transforming or defining systemic anomalies.[69]

Accounting for Taste

The larger public, in which experts were sometimes subsumed, had different ways of making sense of monsters. Anomalies were frequently understood morally or emotionally. Like the hybrids with which they were routinely grouped, monstrosities often inspired their eager audiences with disapprobation and even disgust. For this reason the term itself, especially in its adjectival form *monstrous,* also developed a more general pejorative connotation. It had, indeed, functioned since the Renaissance as a dead metaphor of disparagement, applicable far beyond the realms of natural history, anatomy, or public entertainment.[70]

Monstrous births were frequently explained as punishments divinely visited on erring parents. Thus, it was reported at the end of the seventeenth century, one Mary Blackstone, who "was passionately desirous to have a Child, but God withheld from her the fruit of the Womb," attempted to contravene this dispensation by pharmacological means. Her efforts produced a pregnancy that lasted nearly two years. Ultimately, after "extraordinary pains and labour," she delivered a child with three cyclopean heads, "a small twisting thing in the form of a Serpent about each neck," "the privy Members of Male and Female," and a variety of supernumerary extremities. The disobedient mother died after this ordeal, but not before she had acknowledged her sin in desiring children against God's will and advised other women to avoid the same fault.[71] The pseudonymous author of *The Works of Aristotle, the Famous Philosopher,* which was issued repeatedly throughout the eighteenth and nineteenth centuries, identified a variety of faults for which monstrous progeny constituted predictable retribution, including "filthy and corrupt affection," "bestial and unnatural coition," and the "unclean embraces" of "a man and his wife when her monthly flowings are upon her." The "famous philosopher" acknowledged that, at least in the latter case, the monstrosity of the resulting offspring might not be immediately apparent at birth, but he assured his readers that "children thus begotten, for the most part, [were] dull, heavy, sluggish, and defective in understanding."[72]

Naturalists who shared these feelings were inclined to vent them with more restraint, but the language with which they characterized monsters and speculated about their causes could nevertheless indicate similar disapproval and even revulsion. Charles Lyell described monstrosity as a phenomenon which, if it were not controlled by a set of higher laws, might undermine the entire natural order with its warped and polymorphic creativity. He noted with relief that because "sterility [was] imposed on monsters as well as on hybrids," organic beings were subjected to a process of "perfectibility, in the course of ages," rather than to one of "successive *degeneracy.*" Another fortunate restriction prevented creatures from transcending their appointed place in the taxonomic hierarchy by means of monstrosity: "organized beings may never pass the limits of their own classes to put on the forms of the class above them." A monstrous reptile might therefore resemble a fish, according to this formulation, but a monstrous fish could not resemble a reptile. No matter how many heads a monstrous chicken or lamb might possess, the brains within them could be no more complex than those of more modestly endowed conspecifics.[73] One of the many physicians who wrote in the *Lancet* about monstrous births at which they had assisted expressed analogous aversion with less intellectual distance. In the middle of a dispassionate account of a very difficult delivery, he characterized the dead, deformed, but only mildly monstrous fetus that had occasioned it as "one of the most unsightly objects I had ever seen."[74]

If this excited negativity represented the orthodox moral assessment of monsters, it was far from the only possible reaction. Indeed, its very energy suggested submerged complexities, and again, linguistic usage also indicated alternative evaluations. If the term *monstrous* routinely expressed horror and outrage, in certain contexts it instead connoted admiration and enthusiasm. Colloquial Renaissance speech used it in the sense of "to be marvelled at," which later weakened to a simple intensifier, so that Jonathan Swift could refer casually to a "monstrous deal of snow." In the nineteenth century, the latter usage was replaced by the slang adjective *monster*, appreciatively applied to a range of very ordinary objects.[75]

The attitudes of exhibitors, civic authorities, and audiences toward the display of these theoretically unnerving creatures were similarly relaxed. Their easy acceptance indicated that for many people interest in the anomalous was uncomplicated by strong discomfort, as did also the fact that the striking proliferation of monsters inspired no widespread expressions of alarm. At least among those who paid their sixpences and shillings to gape at monstrosities, grotesque superficial appearance was as little disturbing as subversive philosophical implications. The few showmen who bothered to address the possibility that their exhibits would prove visually shocking felt that the most perfunctory reassurances sufficed to allay audience concerns. For example, potential customers who feared that they would be repelled by the appearance of "A Child . . . with two faces" were informed by the promotional poster that there was "nothing in it unpleasant to the Sight of any Beholder."[76] And even more intense reactions to such ill-favored and potentially disruptive creatures often contained at least as much thrill as horror.

The rhetoric of the promotional literature that ballyhooed individual monsters, often echoed by journalists, transformed sneaking relish into overt enthusiasm. A newspaper report about the two-headed Millie Christine, an American double monster who toured Europe in the 1870s, documented its author's conversion: "We expected to see a monstrosity . . . ; on the contrary, we found her pleasing in appearance, agreeable in her manners, and endowed with good conversational powers."[77] Advertisements for monstrosities of all kinds explicitly attributed similarly appreciative attitudes to the entire range of their audience. A flier announcing the appearance of Sarah Ann Gallant—whose enormous girth (she weighed approximately 120 pounds, although only seven years old) made her an "Extraordinary Living Phenomenon!" and "The Wonder of the Age!"—emphasized that she was one of "those extraordinary Freaks of Nature, that appear . . . to astonish vulgar minds, and lead the learned and scientific to ask, 'Can such things be?'" The handlers of "The Human Tripod, or the Three-Legged Child and First Bipenis ever seen or heard of," addressed their publicity to both "lovers of natural Sciences," who would be inclined to speculate about "the causes of this extraordinary gemination of limbs and organs," and to "the Nobility and Gentry," whose interest would apparently rest on other grounds.[78] Even strictly technical descriptions hinted at pleasurable excitements, as when the *Lancet* referred to an unfortunate West

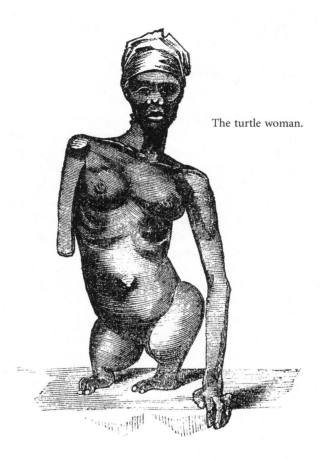

The turtle woman.

Indian with a normal head and torso but shriveled arms and extremely stunted legs, by the midway nickname of "the turtle woman."[79]

Indeed, medical attention to the anatomical grotesque inextricably combined amateur and professional gratifications. Patients sometimes acknowledged the complicated nature of the physician's scrutiny by tacitly redefining their relationship, exchanging the terms of the consulting room for those of the sideshow. For example, a Huntingdonshire doctor complained that the parents of a *"lusus naturae,"* with two cyclopean faces, four arms, one body, and many other abnormalities, at whose stillbirth he had attended, would not allow him to conduct a detailed examination of "the peculiarities of the organization or arrangement of the viscera . . . without a previous money bargain, which I declined."[80] By the end of the nineteenth century, according to two contemporary teratologists, doctors had developed such an "insatiable" taste for monstrosities that "hardly any medical journal is without its rare or 'unique' case."[81]

Proportional Representation

Every departure from the ordinary or natural was thus open to conflicting interpretation. Depending on the beholder, an anomaly might be viewed as embodying a challenge to the established order, whether social, natural, or divine; the containment of that challenge; the incomprehensibility of the creation by human intelligence; or simply the endless and diverting variety of the world. And beholders who agreed on the content of the representation could still disagree strongly about its moral valence—whether it was good or bad, entrancing or disgusting. More fundamental still was the question of whether a given individual was monstrous or not. In the abstract, such problems did not arise. Monsters were tautologically defined as deviations from established norms. But in the concrete world of real monsters, easy binary distinctions were often impossible. Monstrosity had to be determined instead according to a sliding scale of deviation. The resulting graduation redrew the boundaries between the monstrous and the ordinary, making classification a matter of judgment, even of choice.

The most numerous and straightforward anomalies available for admiration and scrutiny were giants, who transcended the natural limitations that restrained the growth of ordinary people and animals.[82] Some of these creatures were distinguished by their girth, like Daniel Lambert, weighing 701 pounds and measuring "Three Yards Four Inches round the Body, and One Yard One Inch round the leg," who chatted affably with his visitors, "many of the first rank and fashion"; or the "Giant Sheep . . . the most wonderful production of Nature, of the enormous Weight of 402 lbs." that toured with a historical panorama featuring "Scriptural, Historical and Tragic Pieces."[83] But the most attractive giants were distinguished by their height, a more majestic dimension than circumference. Their appeal, however, like that of the unusually obese, was essentially statistical. Anyone who approached seven feet could hope to convert his or her physical attributes into cash; anyone who convincingly topped that mark was guaranteed at least passing celebrity.

Most advertisements for giants on display confirmed the absolute and essential nature of their single asset, reductively offering no other enticements than a bald statement of height, provenance, and often age. Height was so compelling that it required no supporting physical assets. Indeed, it was assumed that giants would be ill-proportioned and unattractive. The

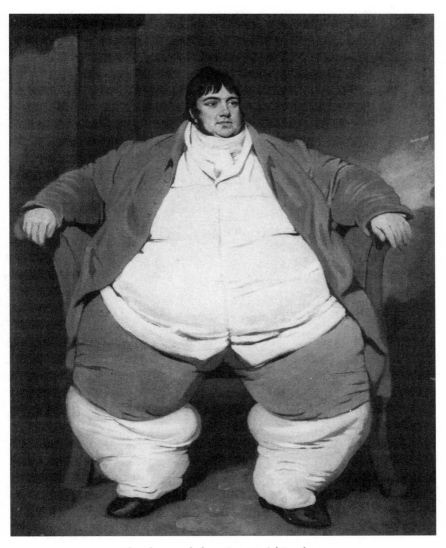

The obese and charming Daniel Lambert.

contrary always occasioned positive notice. For example, the publicist for Anna Swan, the "Nova-Scotian Giantess," claimed that "we have never seen a real Giantess before with anything like so comely a figure," and Charles Waterton recorded that the French giant Monsieur Brice was "remarkably well made."[84] The personalities of giants seldom emerged in either their promotional literature or the accounts of observers.

Spectacular cases of gigantism could prove as attractive to collectors after death as they previously had been to showmen. Indeed, some particularly desirable specimens had to be secured presumptively. Among the most famous giants of the late eighteenth century was Charles Byrne or O'Brien, who toured as "The Irish Giant" and "The Tallest Man in the World."[85] His great height—advertised as eight feet four inches but actually, as revealed by postmortem measurement, perhaps half a foot less—aroused the desire of the medical profession, and especially that of John Hunter. He and his learned competitors correctly anticipated that they would not have to wait long for a chance at his skeleton, although whether he followed "the usual early decline of his class of anomalies" or whether he drank himself to death has been debated.[86] Hunter's most persistent rival in the pursuit of Byrne's skeleton was not, however, a fellow anatomist but the very object of his attentions, who, when he learned of the plans to dedicate his person to science, vowed to resist them. Although he willingly profited from the reductionism of the midway, he was reluctant to undergo the final distillation into a mere specimen. On his deathbed, therefore, he paid some fishermen to take his body to the middle of the Irish Channel and sink it with leaden weights. Hunter over-bribed them, however, and ultimately gained his prize, spending, according to persistent rumor, more than £500 in the process. Byrne's skeleton ultimately assumed a prominent position in the Hunterian Museum, and even the kettle in which the giant's body had been boiled was preserved as a memento and exhibited to the British Medical Association a century later.[87]

Celebrated dwarfs, of whom there were also many, usually seemed more fully human. Only the very young and the very tiny were exhibited as the mute, monodimensional inverses of giants. For example, Caroline Crachami, on show in London in 1824 as the "Sicilian Fairy," was not quite twenty inches tall, having scarcely grown since her birth nine years earlier. Although Crachami could talk, albeit unintelligibly, and communicate her emotions, her admirers, of whom there were more than 200 a day at the height of her popularity, primarily came to observe a passive statistical spectacle. If they paid double the normal shilling entrance fee, they were allowed to handle and examine her. She died suddenly, after a few brief

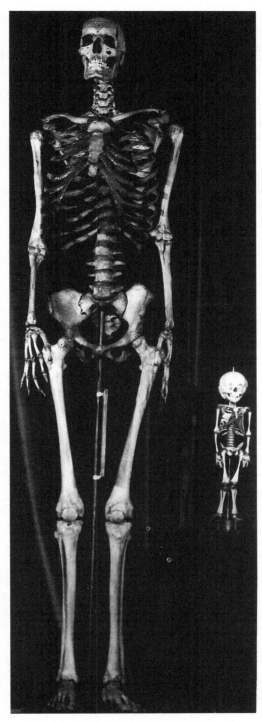

The Irish giant and the Sicilian fairy on display
in the Hunterian Museum.

months of fame, and her fortunes after death resembled those of the Irish Giant. Her parents hoped to preserve her body from its ultimate objectification under the anatomist's knife, but the man who had acted as her impresario—an unscrupulous doctor whom they had first consulted for medical advice—peddled the body to the Hunterian Museum for £10. After dissection, Crachami's minuscule skeleton was mounted next to Byrne's enormous remains. The museum also preserved her shoes, her stockings, and casts of various pieces of her body for posterity.[88]

In contrast to giants, show dwarfs were routinely advertised as admirably proportioned (if undersized) physical specimens.[89] Thus, the 34-inch Mattias Gullia, who appeared in the 1830s as "the Man in Miniature," had allegedly been praised as "one of Nature's most perfect models of Symmetry . . . in all the Academies of France, Italy, and Germany."[90] Dwarfs were expected to be intellectually impressive—a quality that not only enhanced their popular appeal but protected them from some of the worst excesses of midway exploitation—and might display many other accomplishments as well. Nanette Stocker and John Hauptman, exhibited together in 1815, "drew great crowds . . . to witness . . . their wonderful performances . . . on the pianoforte and . . . on the violin."[91] Their social lives, too, seemed noteworthy by virtue of their counterintuitive ordinariness. Dwarfs frequently married, and if their partner was of their own size, they were apt to tour in pairs.

Señor and Señora Santiago de los Santos, who took Britain by storm in 1834, epitomized dwarfish appeal. At 25 inches, Santiago was, as his promotional literature proclaimed, "the smallest man in existence" and therefore "king of all dwarfs"; as it further explained, "when the form of a Man is condensed into the minute form of the smallest Dwarf, just in all human proportions, and evidently stamped with the expression of superior intelligence, the pigmy production rises into an extraordinary miracle." He spoke four languages, danced and sang with great ability, and had manifestly benefited from the "first-rate education" received with the children of the Viceroy of his native Philippine Islands. His wife, who was less remarkable in all respects, being 38 inches high and a native of Birmingham, occupied a smaller share of the limelight. But the narrative of romance and domesticity that she implicitly embodied nevertheless constituted an essential component of the show. One poster mentioned that he had come halfway across the world "to find a Wife." Buried in the small print at the bottom was the coy announcement that "his dear wife . . . is expected to have a little King or Queen."[92]

If the notion that dwarfs lived a lilliputian version of ordinary existence, somehow made more refined and elegant by virtue of its miniaturization, titillated the general public, it could also raise issues particularly troublesome to those whose interest was more specialized. The likely consequences of the domesticity that naturalized dwarfs in the eyes of most onlookers could exaggerate their monstrosity from a systematic point of view. It was reported that the Protestant clergyman who had officiated at the Santiagos' wedding had at first hesitated to marry them, fearing that their union would contravene the canon law prohibition against "propagating a race of dwarfs." His vacillations were ended by the secular legal authority of the high bailiff of Birmingham, who subsequently gave away the bride. They were also retrospectively disparaged by the medical authority of Dr. Henry Davies, who assisted at both of Señora Santiago's labors; he claimed that "however Nature may be now and then fantastic in her productions, she rarely or never perpetuates the same to a second generation."[93]

In support of this contention, Davies offered examples in both genealogical directions: parents of normal height who had produced dwarfs, and dwarf parents whose children towered over them. But counterexamples could not logically eliminate the possibility that prolific dwarfs would give rise to a persistent strain of diminutive people. Such a development would inevitably compromise the distinction between natural and unnatural. Or, to put it more precisely, it would point out a preexisting ambiguity begged by that distinction. After all, human height increased and diminished on an unbroken continuum; as a pseudonymous writer on natural history had pointed out earlier, "we are . . . at a loss to draw that . . . line which should separate the shortest class of mankind from dwarfs."[94] And as dwarfs thus blurred the distinction between the ordinary and the monstrous, they also hinted at still more troublesome taxonomic slippages. Throughout the nineteenth century teratological literature was apt to conflate the dwarfs occasionally born to all groups of people with the uniformly short races that had been reported to live in Africa and other distant places since classical times.[95] This combination of exoticism and monstrosity could seem to violate the very limits of the human, so that the ancients were criticized for overvaluing pygmies, "that diminutive race, which . . . has been long degraded into a class of animals that resemble us but very imperfectly."[96]

Body Parts

Only in their mildest form did abnormalities of deficit spark the casual and sympathetic appreciation that attracted crowds to the Irish giant and the Santiagos. Dramatically missing limbs could occasionally rivet public attention and open the public purse. Early in the nineteenth century a two-year-old child born without arms and therefore, in the view of its doctors, devoid of tactile sensation, was sent from its Kentish home to be exhibited in London. Despite the contrary impression given by its mode of display—and, indeed, by the form of the public announcement—the child was advertised as an object of charity rather than a source of entertainment or enlightenment. The curious and concerned were informed that, since the child had to be taken care of, "the Parent must therefore be supported; and that which the charitable may give, they are requested to note in the book, that those who have taken an interest in the welfare of this Child, may see it protected."[97] Monsters who spectacularly compensated for missing limbs promised patrons more conventional gratification. In the 1840s, for example, the teenaged Miss Spencer, armless and also missing her right thigh, toured as "the most Remarkable Living Phenomenon." Her posters proclaimed that, "notwithstanding her peculiar bodily deformity, she has been so gifted by the Beneficent Creator, that she is able to do all kinds of Work with her Toes!"[98] A "Beautiful High-bred Bright Bay Colt . . . foaled with ONLY THREE LEGS," on show at Portsmouth several decades earlier, was analogously claimed to be "full of Vigour, and walks very nimbly."[99]

Such phenomena could also excite at least modest interest without the advantage of advertising. The Wernerian Natural History Society of Edinburgh devoted an entire meeting in 1822 to "an account of the natural expedients resorted to by a boy in Chesire, for supplying the want . . . [of] both of his fore-arms and hands."[100] In the same year, a poor boy named Jones was apprehended for causing a disturbance on London Bridge. People had congregated to stare at his monstrous extremities, which a broadside characterized, using the language of systematic transgression, as "fins, which occupied the place of arms" and a "hand, at the extremity of the right leg," although "the other parts of the body, and the head, were those of a human being of the male sex." Even the dexterity with which the boy compensated for his deformity, delicately wiping away his tears with his toes, evidenced his disruptive position, simultaneously apelike and human, in the natural order: he "showed as great a command of the leg as those animals do, who become more ugly the nearer they approach in their resemblance to man."[101]

More dramatically deficient monsters, which demonstrated the extreme possibilities of this kind of variation, were less vulnerable to general appreciation. Whether human or nonhuman, most of them were small, dead, and preserved in jars—either nonviable fetuses, stillborn babies, or infants who had survived their births only a short time. Some of the most interesting

"A freak of nature."

deficiencies were not obvious externally, such as the atrophied or missing viscera in a fetus whose abdomen, when opened, turned out to consist mostly of an enormous bladder.[102] And even after they had been revealed by dissection, a procedure which itself lacked widespread appeal, the significance of such anomalies might remain obscure to those who lacked at least rudimentary anatomical knowledge. Finally, many monsters of deficiency, especially those in which the deficit was severe—for example, a preserved specimen displayed to the Westminster Medical Society in 1848, which was more or less ordinary from the waist down but an undifferentiated mass of flesh above, topped by a fringe of hair—seemed repellently and incomprehensibly grotesque, even for the relatively robust and ghoulish taste of the eighteenth- and nineteenth-century public.[103]

If such creatures were meat too strong for the vulgar appetite, however, they exactly suited the refined inclinations of the medical community. The *Lancet*, for example, scoured the world for these phenomena, briefly noticing examples culled from foreign journals to complement the expansive accounts of home-grown monstrosities.[104] These elaborate descriptions were couched in the distancing language of technical anatomy; thus one "foetal subject" presented itself with "urethra nearly imperforate, . . . penis destitute of a corpus cavernosum . . . liver . . . atrophied, and compressed against the diaphragm, . . . [and] the ileum . . . larger in its diameter than the colon," among other afflictions.[105] Nevertheless, the conventional organization of such accounts tacitly insisted on the presence of the recording physician or surgeon. They were usually structured around a double or repeated chronology reflecting the observer's experience, so that the anomaly was first described externally, as it had emerged during parturition, and then again internally as it had been gradually laid bare by the dissecting instruments.[106] The subdued intensity of the authorial voice in these cases, along with the precise detail of both the verbal descriptions and the accompanying illustrations, suggested that the *Lancet*'s medical and surgical readership shared the fascination with the lurid that inspired general interest in monstrosity, only exacerbated by the stronger stomach bestowed by habitual anatomical exploration.

There were also less prurient reasons for their interest. When a doctor encountered an anomaly in the course of his routine professional duties, abstract questions suddenly became startlingly concrete. The requirements of medical ethics might be relaxed in the presence of a monstrous infant—especially if it was considered not fully human. A surgeon who had attended at the birth of an "acephalous monster" admitted that he had "divided the [umbilical] cord without previously tying it"; he asked his colleagues whether he was "justified under the circumstances in doing this?"[107] The possibility, or at least the desirability, of preventing such monstrosities also gave ordinarily theoretical issues a practical edge. Even though, as one contributor assured his more inquisitive colleagues in 1848, "the first cause of these anomalies is yet beyond our knowledge,"[108] it was hard to avoid the question with which a Leicestershire physician plaintively ended a letter describing an "anencephalous monster" that he had delivered: "How are such arrests of normal development to be accounted for?"[109]

That deficient monstrosities were in some sense accidental occurrences was emphasized by the formulaic assertion that their parents had a previously unassailable reproductive track record: for example, a doctor who delivered an infant cyclops in 1862 assured his colleagues that "the father and mother . . . have six other children, all healthy."[110] So the question that practitioners tended to formulate was why the natural order had been so abruptly and spectacularly abrogated. Sometimes the proposed answer conflated the natural and the social. Fathers who abandoned their ordained protective role could thus deform their unborn children. With respect to one "acephalous monster," the doctor observed that "the father of the child had severely beaten the woman's head against the edge of a well when she was two or three months pregnant."[111] The mother of a "monstrous female foetus" had "many times received ill-treatment from her drunken husband."[112] In the absence of abusive treatment or other physical suffering, speculative doctors were apt to fall back, albeit without full confidence, on the mother's mental impressions. Thus the armless Jones had been born shortly after his mother "heard of the loss of her husband at sea; and . . . his deformity was, by the physician, attributed to the fright"; for the birth of a cyclops whose parents were "of highly nervous temperament—the mother hysterical," the attending doctor could offer no other explanation than the mother's exposure to both a harelip and a cleft palate in the early stages of her pregnancy.[113]

Monstrosities whose endowments were unusually generous presented the unnatural in a less alarming light. They figured more frequently in public exhibitions; indeed, sideshows and museums overflowed with animals and people sporting extra limbs. Although in an earlier age the birth of such anomalies even to farmyard livestock had seemed portentous, as they grew commonplace they became merely diverting, almost as much an extension of the natural order as a violation of it.[114] The anachronistic aversion lingering in some rural districts occasionally doomed perfectly viable monsters to a premature grave; for example, the *Annals of Sporting and Fancy Gazette* reported in 1822 that a "strong and healthy" kitten with a double face, born in Dumfriesshire, had been drowned because of "the superstitious notions of its owners."[115] As the censorious tone of the reporter implied, however, the more usual impulse was to cherish such creatures. The same journal noted regretfully a few years later of a calf with two "beautifully formed" heads that, "had it not been neglected the night it was calved, no doubt is entertained but it would have lived."[116] A litter of monstrous kittens—one with two heads, one with six legs and a double spine, and one of a liver color allegedly unprecedented in felines—was treated as a charming, if grotesque, nativity scene, attracting "a great many respectable people . . . to see *the lady in the straw*."[117] Nineteenth-century advertisements for such attractions used language that promised delight unleavened by awe or terror. A "six footed Heifer with five legs" was "a beautiful Beast," a ram with six legs was "curious," a two-headed heifer was "surprising."[118]

The most provocative double monsters were those whose superfluity seemed essential rather than peripheral. Supernumerary legs and arms could be considered merely extra, only somewhat more striking than the additional fingers and toes that constituted such trivial departures from the norm as to be discussed under the rubric "hereditary formations," rather than "deformations" or even "malformations."[119] The juncture of entire bodies, on the other hand, implicitly undermined the distinction between singular and plural by breaching the boundaries that naturally defined and isolated each individual. It thus potentially extended the monstrous to the mental or spiritual plane. One account of Chang and Eng, the eponymous Siamese twins, whose show career extended through the middle of the nineteenth century, began with the queries: "Is it one *man* in two bodies . . . ? Is there but one life, one responsible will . . . only with double bodies . . . ? Or, are they really two *men*—each as distinct from the other as Smith from Brown, with the exception of that one mysterious bond?"[120] Interest in these issues was not limited to naturalists or philosophers. Chang and Eng were exhibited all over the world and inspired a literary outpouring described as "simply stupendous" in amount and variety.[121] As if to emphasize the breadth of their appeal, as well as to suggest that useful light could be thrown on the central question by commonsensical as well as expert analysis, published discussion of double monsters often intertwined anatomical description with anecdotal narrative. A pamphlet featuring Chang and Eng included a chatty biography, a soberer recapitulation of the same material by the professor of medicine and midwifery at the University of

Double pig on show.

THE SIAMESE TWINS.

CHANG AND ENG.

Edinburgh, a discussion of the possibility of separating them by a distinguished surgeon, and illustrations of their recreations and their wit.[122]

Although the organic doubleness of such monsters was the symptom of the ultimate problem that they posed, its solution did not lie in mere anatomy. The scalpel was unlikely to reveal the relation of the personality to the person—of the soul to the body. As a result, if conjoined twins were born dead, they were apt to receive perfunctory technical descriptions in specialist journals and then to be preserved, if their parents were willing, for the edification of future medical students.[123] As one contributor to the *Lancet* put it, "to theorise upon their separate existence, and, had they lived, moral responsiblities, would be but to wander in the mazes of romance."[124] But if such monstrosities survived, they began immediately to reward a different kind of attention. The author of "A True Relation of a Monstrous Female Child" (joined twins born in Somerset in 1680) asserted—long before they could have persuaded him by their conversation—that they possessed two immortal souls.[125] Similarly, a two-headed monster born in Sicily in 1829 (given the standard somatic synecdoche for intelligence and will, two heads were equivalent to two entire bodies) at only nineteen days old gave some evidence of double autonomy, since "the children . . . appear to have the sensation of hunger at different times."[126] The author of the report indicated that this evidence did not seem entirely conclusive by alternating between the pronouns *it* and *they* in describing the monster. The parents expressed their firmer conviction by giving each head her own name, although they exhibited them in Paris under the composite of Ritta-Christina.[127]

Ritta-Christina survived only a brief time on show before succumbing to chilly Parisian interiors, but from time to time double monsters attained adulthood.[128] Their maturity provided more voluminous but never completely decisive testimony. Some of this evidence was anatomical and physiological. Even if doubleness of body did not guarantee doubleness of mind, the extent of duplication, especially with regard to the nervous system, was suggestive. Chang and Eng, who were united at the chest, seemed to possess only a single set of sense organs between them—each had one good eye and one good ear—while some other functions were completely independent. For example, their urinalyses proved no more similar than would those from any two unconnected people.[129] Millie Christine, bonded at the back of the pelvis but otherwise double, shared sensation below the point of union but felt separately above. According to the professors of medicine who examined her or them, each head possessed "separate intellectual

BIOGRAPHICAL SKETCH
OF
MILLIE CHRISTINE,
THE TWO-HEADED NIGHTINGALE.

"None like me since the days of Eve—
None such perhaps will ever live."

faculties, . . . and the volitions of the will are independent, but very much in harmony with each other."[130]

There was, of course, a sense in which the question was meaningless. It was clear that individuals whose lives were so ineluctably united could not afford much independence, even if they were capable of it. This point was humorously made by the widely circulated but apocryphal story that Chang and Eng, who ultimately settled in the United States, had had their first serious quarrel at the outbreak of the Civil War, one siding with the North and one with the South. In fact, their lives were, apparently, models of concord. Not only did they get along with each other, sharing the same opinions as well, perforce, as the same vocations and avocations, but each was married and the father of nine children. The families, less harmonious than their progenitors, lived in separate houses and were visited in rotation by the twins.[131] Millie Christine was likewise at peace with herself or herselves. According to one promotional biography, "the tastes and habits of the two are precisely similar."[132]

Vive la différence

Among the varied kinds of monstrosities, one presented special taxonomic difficulties. Although hermaphrodites incontestably diverged from the norms established by both their parents and their species (at least if they were vertebrates), they did so in a way that made it particularly difficult to relate them to other anomalous creatures. When John Cleland, undeterred by the taxonomic failures of his predecessors, attempted his own classification of "monstrous and anomalous forms," he followed Hunter by dividing them, in the first instance, into those of defect and those of excess. He specifically excluded hermaphrodites from this neat binary arrangement, however, on the grounds that "true hermaphrodism, the formation of both ova and spermatozoa in the one individual, is neither anomaly by defect nor by excess, but is altogether *sui generis*."[133] And begging this systematic question merely introduced others. Hermaphrodites varied widely among themselves, in ways that similarly resisted orderly categorization. In 1839, for example, an obstetrical expert sharply divided the "two distinct varieties of hermaphroditic malformation": the *"Spurious"* (where the organs of one sex merely mimicked those of the other) and the *"True"* ("cases in which there is an actual mixture . . . upon the same individual").[134] On the other hand, Robert Knox insisted that these ambiguous beings formed a series so subtly graduated as to preclude clear differentiation not only between the various types of hermaphrodites but also between hermaphrodites and incontestable males and females. Knox asserted that in humans, as well as in pigs and cattle, organs characteristic of one sex were frequently present in the other, but that only when such organs became "effectual in modifying and altering the whole economy of the being—the appearances are . . . deemed worthy of notice."[135]

Knox's characterization of the interest aroused by manifest hermaphrodites was, if anything, an understatement. Creatures who presented evidence of belonging to more than one sex evoked predictable, intense, and unflagging fascination. An eighteenth-century critic of the Royal Society alleged that "Histories of Hermaphrodites have long had a first place" among "miraculous Accounts of Things out of the Course of Nature," and that the distinguished savants were as eager and gullible as "People, unacquainted with the Structure of a human Body, and the Laws of Nature in its formation."[136] A century and a half after the Royal Society first contemplated ambiguous sex, the minutes of the Wernerian Natural History Society of Edinburgh emphasized that a presentation had discussed "Congenital Anomalies, including hermaphrodites."[137] They, or at least their superabundant reproductive organs, were prized additions to private and public collections. At different times, for example, correspondents let Joseph Banks know that a "Hermaphrodite sheep" and "the two Heifers which we suspected to be imperfect Hermaphrodites" were ready for disposal.[138] Among the "Specimens of Comparative Anatomy" that Everard Home listed for the Hunterian Museum conservator in 1802 were the "Generative Organs of An Hermaphrodite Cod having the soft and hard Roe in the same fish," the "Testicles of an hermaphrodite Dog," and the "Clitoris of the same Dog"; a catalogue of the same collection published three decades later listed a whole series of "Hermaphroditical Malformations."[139] In 1851 the Anatomical Museum of the University of Glasgow gratefully recorded the receipt of the "Genital organs of an apparently hermaphrodite goat."[140]

Students of anatomy unable to examine these ambiguous remains in person could read about them in the specialist press. In 1838 the Medical Society of London viewed "a preparation of malformation of the genital organs" taken from a country lady who, it was asserted, "had always been of a reserved character." A bladder affliction shortly before her death had allowed her doctor to suspect that, as one appreciative observer put it, "this case came nearer to . . . exhibiting the rudiments of both sexes than any he had . . . seen." The account of the meeting in the *Lancet* included a detailed description of her organic irregularities, along with, for comparison, those of a "young woman, who . . . bore every resemblance, in conformation, to a fat boy" and had "got married merely because she had known her husband a long time, and wished to cook his victuals, &c." and those of "a well-formed man, with a beard and other characteristics of manhood" who had "married a beautiful woman three years since, and had never made the slightest approaches to her, for reasons which his wife did not know."[141] Reports of similarly "interesting patients," such as John Battle, who showed "deficiency of masculine character," having been admitted to Charing Cross under the name of "Elizabeth" and wearing a dress, flowed in from hospitals and other institutions, while private practitioners submitted accounts of infants whose anatomy made it "difficult to ascertain whether these organs belonged to a male or a female child," often requesting collegial advice about the official sex designation.[142] An autopsy of a juvenile orangutan who died in New York of "excessive indulgence in fruits" revealed that the animal, already "interesting" on account of its species, furnished "the nearest approach to a complete union of the sexes in the same individual which has been detailed" (the closest competitors cited were a French bull with ovaries and uterus and a Portuguese woman with external—albeit "anatomically" unexamined—male genitalia). The detailed anatomical description of the hermaphrodite orangutan was accompanied by drawings of the internal and external reproductive organs, as well as by learned musings about the "most interesting results [that] might have been elicited" had the orangutan lived to maturity: for example, "by masturbation, might not the animal have impregnated itself?"[143]

By considering the synecdochic genitalia of hermaphrodites independently of the rest of their dismembered bodies, which were often similarly ambiguous, anatomists may have seemed to remove them from vulgar curiosity—or at least to emphasize the rarefied character of their own intensive examinations. But if the appeal of these preserved and partial specimens was limited, when hermaphrodites were exhibited alive and entire, everyone wanted to see them. Like other spectacular monstrosities, they drew miscellaneous crowds of people who paid their shillings or sixpences for the privilege of gaping, although both observers' accounts of hermaphrodites on display and the brief narratives included in advertisements often suggested that their subjects could be apprehended on two levels. Roughly speaking, general admission entitled the audience to admire the mingling of secondary sexual characteristics, while examination of the primary sexual characteristics was restricted to those who demonstrated more serious interest, in cash or in some other coin.

A manuscript illustration of such a display, dating from the beginning of the eighteenth century, showed the hermaphrodite elaborately clothed, but a flap over the groin could be lifted to reveal his or her nakedness.[144] The broadside advertisement for the "Famous African," exhibited in London and the English provinces in the 1730s, made the same point linguistically. It proclaimed the combined masculine and feminine nature of facial features, voice, chest, musculature, and formation of arms, hands, and thighs in plain English, and then referred, "For a particular Description of the *Parts which distinguish the Sexes* . . . to the short Account in *Latin,* following." Those lacking classical education had to be satisfied with the separate assurances of three distinguished doctors that this "Wonder of the World" was *"perfectly Female," "perfectly Male,"* and "neither Sex perfect, but *a wonderful Mixture of both."*[145] On a "brief visit to Dieppe" over a century later, a correspondent of the *Lancet* reported that he had seen a girl with extra legs "exhibited in an itinerant show by her father and mother to all comers, and of course clothed, for a few *sous;* but I found no difficulty in obtaining a *séance particulière* for the moderate fee of a couple of francs, at which I had full opportunity of examining this curious *lusus naturae.*" He discovered that the child's doubleness extended into the pelvis, so that "the development of the genital organs" was "most remarkable," including, as far as he could tell without vivisecting, a complete female set and an incomplete male set.[146]

Shared concern with sexual abnormality nevertheless cut across the distinctions of class, education, and temperament that otherwise divided the audience for monstrosities. After all, the borderline most spectacularly compromised by hermaphrodites was constructed and defended by social rather than scientific taxonomy. Although maleness and femaleness were striking facts of nature, and reproduction was the focus of a great deal of zoological attention in eighteenth- and nineteenth-century Britain and elsewhere, sex distinctions were not ordinarily used to identify species, genera, or more comprehensive taxa. On the contrary, they figured in classification primarily as the occasion for fanciful speculation or as the source of mistakes. As John Fleming rather irritably explained, "the male is considered as the representative of the species . . . When a female individual comes under notice, it is frequently very difficult, if not impossible, to determine the species to which she belongs."[147]

Indeed, the language of taxonomic discrimination could function as a metaphor for preexisting convictions about the great gap that separated the sexes. The frequency of mild sexual dimorphism in animals, along with the assumption that sexual difference was likely to make itself obvious, meant that animals that resembled each other, such as the quagga and the zebra, were sometimes confounded as male and female of the same species.[148] Conversely, gross sexual dimorphism occasionally led naturalists to assign males and females of the same kind to different taxa. Meditating in 1840 on "the whole Physiological Circle of these great Sea-dragons" (the long-extinct ichthyosaurs and plesiosaurs), Thomas Hawkins confessed himself "haunted by an idea . . . that the Strongylostinus . . . is none other than the male Oligostinus." [149]

In a culture that assumed marked disparity between the sexes, the social disruptiveness of hermaphrodites was both more striking and more troublesome than their violation of anatomical convention. In lectures on medical jurisprudence offered at the University of London, Professor A. T. Thomson cited the case (albeit not a recent one) of an apparent female who had impregnated an unwitting young woman with whom, as was not unusual among members of the same household, she or he shared a bed. The consequence of this transgression was a sentence of death by burning. Thomson was condescending toward such "barbarous treatment," associating it with the canon law, which denounced hermaphrodites as "persons to whom the gates of dignity cannot be opened." He cited Coke to prove the superior generosity of English common law, which specifically allowed hermaphrodites to inherit property, "either as a male or a female, according to that kind of sex which doth prevail." [150] Of course, the potential disruption caused by propertied hermaphrodites was less than that caused by hermaphrodites who contracted marital obligations that they could not fulfill, or by hermaphrodites whose sexual activity took paths unpredictable to their close associates.

Even the appropriate packaging and labeling of their sexuality could not entirely meet the challenge that hermaphrodites presented to order, if not necessarily to law, by their very existence. As Knox put it at the end of his very technical anatomical analysis, "hermaphroditism . . . is a condition, as applied to man, opposed to all the existing states of social intercourse . . . the whole frame of human society hinges upon a complete disunion of the sexes." [151] The experience of individual hermaphrodites showed that, no matter what theoretical provision might exist for them, people who blurred that borderline could not be accommodated by the society into which they had been born. Their lives were usually marked by secrecy and confusion. If an infant's mixed endowment was obvious at birth, the parents were apt, as was the case with a "famous Hermaphrodite" exhibited in London in 1750, to "conceal her Deformity"; they gave her a girl's name and brought her up as a girl. [152] If, however, a hitherto female (or at least nonmale) child suddenly discovered an unsuspected masculine dimension, which could happen at the onset of puberty or as a result of some external physical injury, the newly ambiguous girl or boy might find it difficult to fill the gender role to which she or he had been brought up.

But definitive alteration was also problematic. There is the example of a young man who, "up to the age of twenty years," had "considered himself . . . a girl" and had dispensed sexual favors under that impression but was informed to the contrary by a surgeon who treated him for an inguinal hernia. Despite his persistently feminine appearance, he embraced his new identity with enthusiasm and attempted to copulate as a male. This experiment proved unsuccessful, however, and "he was obliged, much against his pride, to return to his old habits and former acquaintance."[153] The stories of other individuals of ambiguous sex suggested that inability to adapt definitively to conventional categories could produce still more destructive psychological consequences. An old woman who died in Guy's Hospital in 1829 attracted attention because of her unusual "robustness of frame." The autopsy showed imperfect male reproductive organs, and inquiry revealed she had led much of her adult life as a man, working as a brickmaker, a groom, and a milkman before settling down to keep a greengrocer's shop. It was reported that "her habits and manners were rude and bold, sometimes indicating a degree of derangement; more than once she engaged with success in pugilistic encounters."[154]

Maintaining female identity under such circumstances could be at least equally troublesome. One Jane W., whose behavior in the lunatic ward of the Macclesfield Workhouse in 1856 immediately made the nurse suspect that she "was *not a woman*," not only annoyed the young women in her ward with her "amatory propensities" but more generally displayed "very depraved" tastes, "as she will eat old poultices with great delight."[155] Nor did social ostracism require such flamboyant eccentricity. In 1842, the household of a country gentleman was thrown into an uproar when a housemaid complained that the cook Mary Anne, who slept in the same bed, took "strange liberties . . . with her person." Other suspicious circumstances, such as the cook's solitary habits, and the fact that "she was never seen to comb her hair, nor to wash her person, nor to allude to circumstances which females generally communicate to each other," convinced her fellow servants that she was a disguised man. So loathsome did they find this idea that they refused to touch any food she had prepared, confining themselves to bread and milk, which they "considered uncontaminated by the touch of abomination." Mary Anne's external appearance was perfectly female, if somewhat hirsute, and she expressed no doubts about the gender in which she had been raised. Nevertheless, medical examination revealed that she was, instead, an "imperfect male."[156] Consequently, as the doctor who made the discovery put it, she had to confront an urgent question: "What now to do . . . ? She was expelled, of course, from the *gynaeceum,* and yet scarce worthy to assume the . . . *toga virilis.*"[157]

Unpalatable as it might have been for the retiring or genteel, a life of public exhibition solved several of the problems that inevitably confronted people who had been labeled as hermaphrodites, whether true or spurious. Exhibition provided a workplace unlikely to foster the intense aversion demonstrated by Mary Anne's colleagues for people whose mixed or ambiguous sex had attracted notice. Such notice became increasingly likely with approaching maturity, which also brought the expectation of financial self-support. Thus the parents of the "famous Hermaphrodite" of 1750 had ultimately abandoned their attempts at concealment "when they were advised to make a Shew of her in order to get Money."[158] Their resolution, like similar decisions subsequently made by other ambiguous individuals, had rewards extending far beyond the economic realm. The willingness of hermaphrodites to put themselves on display ironically served to reintegrate them into the social order. When they attempted to participate in ordinary life, even the workaday part of it that did not routinely require sexual performance, they implicitly presented themselves as normal. Once they were displayed as monstrosities, however, the effect of their anomaly was transformed. As public spectacles—manifest and acknowledged exceptions—their presence ceased to be subversive and disruptive, and on the contrary worked to confirm the very boundaries that they would otherwise have seemed to challenge.

At the same time that these displays confirmed the existence of the borderline between male and female, they also emphasized the importance of maintaining the distinction. They were not alone in making this point. Truly spectacular and persuasive hermaphrodites were not particularly common; their frequency was more on the order of Siamese twins than of dwarfs and giants.[159] So fraught with significance and anxiety was the division between male and female, however, that any violation of it was liable to attract at least a little attention; the rubric of hermaphrodism was consequently stretched to accommodate any phenomenon that so much as suggested gender confusion. Even ambiguous aesthetic preferences invited commentary. Thus a Victorian agricultural writer disapprovingly noted that "the Ayrshire farmers prefer their dairy bulls to possess the feminine aspect in their heads, necks, and fore-quarters."[160] Inappropriate behavior was apt to inspire criticism, whether the actors were humans who cross-dressed at masked balls or ambiguous fowls, such as a crowing hen or a cock who, "to make up the loss of the mother to the brood, . . . devot[ed] himself entirely to . . . their wants." Although in the nineteenth century such misguided birds were not ordinarily burnt alive, as had been their fate in earlier times, they were still castigated in a zoological treatise as "instances of monstrosity."[161]

A variety of physical oddities or anomalies were also exploited under the rubric of mixed sex, whether or not any actual or alleged mingling was involved. Any failure to realize full sexual dichotomization could be read as categorical conflation. The effects of castration or neutering were often described in this way, even though that operation produced creatures with deficient rather than redundant reproductive systems. William Yarrell reported to the Zoological Club of the Linnean Society that, in the case of pheasants, "by deprivation of the sexual organs both males and females . . . assumed characters decidedly intermediate between the sexes." Alexander Walker assured a wider audience that neutered men "have a configuration and habits very analogous to those of women," and that "the results of the operation in women are generally the reverse of those which occur to men."[162]

A phenomenon that occurred with some frequency in domestic cattle was similarly pulled within the orbit of hermaphrodism. When a cow gave birth to twins of different sexes, the female calf, called a "free-martin," usually grew up to be sterile and relatively unfeminine in appearance, both externally and internally.[163] According to John Hunter, the free-martin was simply "an hermaphrodite, in no respect different from other hermaphrodites."[164] His conviction was echoed well into the next century. Busick Harwood treated free-martins immediately after "*Hermaphrodites* in general" in his Cambridge natural history lectures; James Y. Simpson included them in his encyclopedia entry on hermaphrodites, noting that "the free-martin does not present an exact analogy in form either with the bull or cow, but exhibits a set of characters intermediate between both, and more nearly resembling those of the ox and of the spayed heifer"; and the mid-Victorian George Vasey, in his authoritative *Monograph on the Genus Bos*, respectfully claimed that "Mr. Hunter . . . has satisfactorily shown that their incapacity to breed, and all their other peculiarities, result from their having the generative organs of both sexes combined."[165] Despite the near universality of these convictions, a small but not trivial number of free-martins (just under 10 percent) did become pregnant and produce unexceptionable offspring; indeed, one veteran Victorian cattleman thanked such a mother for some of his "finest females."[166] The classification of free-martins as sterile hermaphrodites meant that these unusual but predictable occurrences were greeted as amazing breaches of the laws of nature. Correspondents of the *Farmer's Journal* urged the owner of a reported freemartin in calf to have the birth "effectually certified . . . by the testimony of respectable neighbours, and this verified by some of the nearest Magistrates or Clergy, or both."[167]

So persistently compelling was any whiff of hermaphrodism that anomalies sufficiently striking—even monstrous—in their own terms were presented as violations of the boundary separating the sexes. The death of fifteen-year-old Thomas Lane in 1814 was attributed by the attending physicians to a "foetus" found in his abdomen, even though their report emphasized his incontestable masculinity and explained the unfortunate phenomenon as a partly developed twin, rather than "the fruits of an unnatural crime."[168] A cast of the fatal object was labeled "Foetus abdominalis from a Boy 16 years of age" in Joshua Brookes's museum, and the lurid connotations of the word *foetus* proved so much more powerful than the relatively prosaic and commonsensical commentary in which it was embedded that fifteen years later a lecturer at Guy's Hospital asserted that Lane had "literally, and without evasion, . . . [been] with child."[169] Although the component parts of most double monsters, like Chang and Eng or Millie Christine, shared the same sex, this was not always the case. The rare exceptions were referred to as "of an opposite or hermaphroditic sexual type."[170] The publicity for Lalloo, an Indian teenager who toured Victorian Britain as "the Greatest Living Wonder on Earth," made detailed reference to his "one perfect head," "four perfect arms," "four perfect hands," "four perfect legs," and "four perfect feet," as a preparation for the more discreetly indicated but still climactic attraction: "in fact it is simply a Boy and Girl Joined Together."[171]

Bearded women were similarly appreciated as incipient hermaphrodites, despite the fact that, as one late-nineteenth-century commentator noted, "extreme hairiness in human beings is by no means singular" or limited by sex; he could remember whole hairy families on show in London.[172] For example, the Mexican Julia Pastrana, the most celebrated Victorian example of this type, was both displayed and described in exaggeratedly feminine terms that tacitly emphasized the contrary expectations aroused by her luxuriant beard. One publicity poster asserted that "her eyes are large and fine, . . . her form and limbs, are quite perfect, with wonderfully small hands and feet"; she also possessed a limited range of distinctively feminine accomplishments, notably dancing and singing.[173] The *Lancet* made the same point more directly, noting that "her breasts are remarkably full and well-developed" and that "she menstruates regularly."[174] Ultimately Pastrana offered incontrovertible evidence of her femininity by producing a hairy son, but neither one of them long survived the delivery. The charisma of her hirsutism was, however, apparently undiminished by this change of condition. Her husband/manager had both mother and child embalmed, and, accompanied by a second bearded wife, continued to tour with them for several decades.[175]

If the boundary between the sexes was mainly an intraspecific concern— albeit one that was highly charged and therefore intensely fascinating—the fact of its existence had broader taxonomic ramifications. As individuals of apparently mixed sex threatened to unsettle human gender categories, they could also undermine the equally well-defended barrier that separated

Lalloo: all boy and half girl.

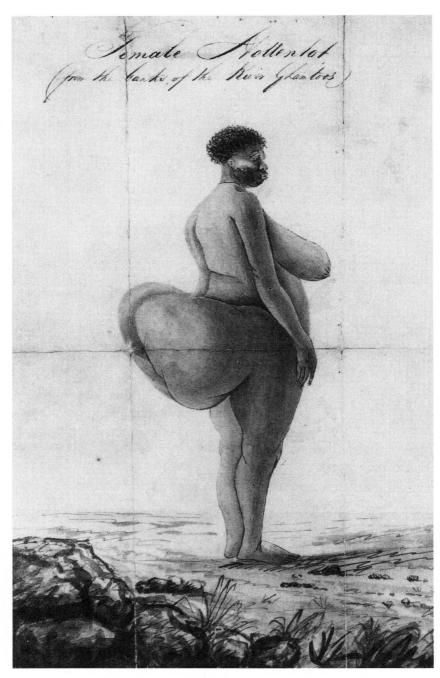

Female Hottentot
(from the banks of the River Ghantoos)

A bearded and steatopygous Hottentot: exoticism compounded
by a hint of hermaphrodism.

human beings from the rest of the animate creation. It was well recognized that hermaphrodism occurred routinely in some organisms, but only in plants or in such animals as snails and worms, which seemed very different from people. A few mammalian species were repeatedly suspected of hermaphroditic tendencies, but such suspicions were readily dismissed as misunderstandings of cryptic superficial evidence. For example, in 1823 the *Annals of Sporting and Fancy Gazette* scouted the "notion . . . of *hermaphrodite hares,* and that these animals are alternately male and female," attributing this "strange conjecture" to "the formation of the genital parts of the male."[176] Thomas Bewick observed that "what has been fabulously related concerning the uncertainty of sex in the Hyena, appears much more strongly to the Civet; for in the male nothing appears externally but three apertures, so perfectly similar to those of the female, that it is impossible to distinguish the sex otherwise than by dissection."[177] Eighteenth-century zoologists consequently named the palm civet of Southeast Asia *Paradoxurus*

The bearded but feminine Julia Pastrana.

hermaphroditus, despite the irreproachable conventionality of its reproductive practices.[178]

As often happened, recognition of difference led inexorably to ranking. John Hunter explained that "the natural hermaphrodite belongs to the inferior . . . genera of animals"; as he further elaborated, "all the most perfect animals are of two sexes . . . The chief distinction . . . is in the parts of generation: but . . . the male may be always distinguished by his noble, masculine, and beautiful figure . . . [especially] in fowls."[179] Most later authorities followed Hunter's lead, asserting, for example, that "amongst the more perfect animals, at the head of which man is placed, the sexes are always found in separate individuals," or more modestly that "the disposition to hermaphrodism is more rare as we advance in the scale of perfection, or, rather, to a more complex organization."[180] Conversely, the "most perfect hermaphrodites . . . are only found in the lowest classes of animals."[181] And, according to one late-nineteenth-century French specialist in reproduction, this hierarchy was also replicated in the vegetable kingdom, where "distinct separation of the sex . . . is a mark of their height in the scale, and of their splendor." [182]

Even a hermaphrodite enthusiast like Robert Knox tacitly acknowledged the existence of this scale, at the same time that he rejected its extreme logical consequences: "It may be supposed by some, that hermaphroditical appearances . . . are limited to animals which are usually considered as lower in the scale of animal creation; but this of course is a great error."[183] More frequently, however, commentators embraced these consequences, asking whether there was "really ever an hermaphrodite in a being so elevated in the scale of nature as man?" or simply asserting the impossibility of such an anomaly.[184] At least until the end of the nineteenth century, when investigators began to up the ante, demanding that hermaphrodites possess both ovaries and testes rather than more miscellaneous evidence of mixed sex, such declarations had little apparent effect on either the supply of hermaphrodites (a steady trickle) or the fascination that they exerted. But they did exacerbate the difficulty of the position in which ambiguous individuals found themselves, intensifying and complicating their monstrosity. If mingled reproductive organs indicated taxonomic inferiority, then human hermaphrodites were not merely transgressive but defective and degenerate—even, from an evolutionary point of view, atavistic. Not only did their anomalous sex exclude them from the social order, but it made them hard to fit into the order Primates (or Bimana) and the class Mammalia.

The Plausible Impossible

Dwarfs and giants, Siamese twins and bearded ladies, however unusual and striking, could be viewed as exaggerations of what was normal or ordinary. They pushed the limits of the natural without threatening to overturn them completely. In general, the questions they posed involved location in the scheme of things: where—and how well—they fit into an ordered cosmos. Alongside the stuffed and bottled remains of hermaphrodites and eight-legged cats, however, were exhibited anomalies that broached more serious versions of these questions or more serious questions altogether. Their systematic eccentricity seemed so pronounced as possibly to require not mere modification of previously recognized laws but wholesale rewriting.

Before the beginning of the eighteenth century, creatures suspected to be imaginary had come to bear a different relation to the natural order than those that were merely monstrous, although they had shared much of their previous history, as documented in many of the same classical, medieval, and Renaissance sources. This divergence did not diminish either the supply of such conversation pieces or the demand to see them. For example, fish furred against the cold were first reported to swim in Canadian waters by a puckish seventeenth-century Scot, who then confirmed his surprising observation by sending a stuffed specimen home; contemporary and later naturalists similarly documented and displayed ancestors of the modern jackalope or horned hare.[185] If these factitious monsters physically resembled the curiosities of an earlier age, however, their meaning had definitively altered. As the parameters within which nature was conceived to operate hardened, the range of the possible correspondingly contracted. Although the categorical transgressions exemplified by such patchwork creatures had been equably encompassed by the capacious vision of an Aldrovandi, his more methodical and systematic successors found them impossible to accommodate. Indeed, the accurate identification and classification of monsters could be regarded as a test of zoological competence. The genuinely liminal could be coopted, but the spuriously liminal required vigilant exposure and rejection. Mere explanation was no longer an adequate response; bizarre specimens had to be explained away.

As a result of this imperative, experts parted company with many who shared their fascination with the unusual. In the presence of the potentially impossible, curiosity and credulity, which traveled together through many museums, sideshows, and tavern back rooms, could suddenly become incompatible. Thus, the intransigent classificatory problems posed by puzzling creatures also produced or mirrored taxonomic issues within their audience. Not everybody felt compelled to eliminate the impossible. Indeed many— including entrepreneurs, their eager audiences, and the opportunistic press—reveled in it, resisting efforts to deprive the world of its magic. This resistance, however motivated, constituted at least a tacit challenge to the definition of the physical world not only as the realm where natural laws operated, but also as the realm where specialists held interpretive sway. As the counterfactual creatures undermined the systematic structures erected by naturalists, so their defenders contested previously staked out scientific turf. And the human taxonomy generated by the recrudescence of myth was equally unsettling. It did not always clearly distinguish those within the institution of natural history from those outside it. That is, some people with claims to expertise were nevertheless inclined to defend rather than to debunk.

This territorial imperative may help explain the lively and persistent interest of skeptical and serious observers in creatures of this kind—creatures that they might otherwise have been expected to ignore. Not even physical evidence was necessarily required to engage the earnest consideration of the learned; when the source was especially authoritative, hearsay offered sufficient inducement. Although well attested in legend and in patriotic British iconography, for example, unicorns had left few tangible remains. It was true that a "long, white, wreathed, Ivory-like Body" identified as the horn of a unicorn was "preserved in the Musaeums of all the Collectors"—including, early in the eighteenth century, that of the Anatomy School at Oxford. At mid-century, however, such objects could be confidently identified as the tusks of the narwhal (a small arctic cetacean); naturalists who continued to attribute them to imaginary terrestrial beasts could be confidently identified as "ignorant" and "absurd."[186] As the author of a late-eighteenth-century compendium fair-mindedly summarized the unicorn situation: "whether this animal ever existed . . . , we are now scarce able to tell, since there is no living testimony of its existence, nor has been for several ages."[187] No subsequent discoveries redeemed the evidentiary status of these once-prized mementos. By the end of the Victorian period "unicorns' horns" had come, in the eyes of progressive modern curators, to symbolize the dark ages of museology.[188]

Nevertheless, unicorns continued to figure in zoological discussion throughout this period, and not always, in Alice's reciprocated characterization, as "fabulous monsters."[189] In part this was because the unicorn possessed a kind of theological legitimacy. It was easy enough to dismiss the testimony of pre-modern bestiarists and mythographers, especially since, as William Frederic Martyn noted in his encyclopedic account of late-eighteenth-century natural history, "the ancients themselves appear to have questioned the existence of this creature."[190] But scriptural attestations demanded more respect. The King James version of the Old Testament referred repeatedly to the unicorn.[191] And if the precise nature of biblical unicorns still remained unclear, their existence, in some form or other, was therefore considered beyond question by the devout. The minimalist response was to assume a mistranslation, and so acknowledge the authority of modern science without outraging that of traditional religion. Commonsensical naturalists therefore argued that the animal in question must be known to zoology under some other, more familiar name. For example, an early guide to the London zoo pronounced that "the Unicorn, as is perfectly beyond controversy, is the Rhinoceros." According to the Naturalist's Library volume on deer and antelopes, the oryx family was "remarkable, as it is supposed that from some of its members the far-famed Unicorn would be made out."[192]

There were, of course, obvious practical objections to these hypotheses. The silhouette of the rhinoceros differed strikingly from that conventionally attributed to the unicorn, and its horn, though single, was short, stubby, and located on its snout rather than its forehead; the oryx came closer in general conformation, but sported two horns. But undeterred by the derision of skeptics, many investigators clung to the unicorns of legend. Rather than dismissing unreliable early reports or attempting to account for them in terms of prosaic reality, they searched and argued for a fabulous animal hitherto unknown to science. In 1886 Charles Gould summarized the credo of such special pleaders, "I find it impossible to believe that a creature whose existence has been affirmed by so many authors, at so many different dates, and from so many different countries, can be merely . . . myth."[193]

Wherever they roved, European explorers therefore kept a sharp eye out for unicorns. John Barrow sought them during his arduous travels through southern Africa in 1797 and 1798, although without success.[194] A few decades later, a natural history popularizer announced that although "the existence of such an animal as the unicorn has long been considered as purely fabulous, . . . very recent accounts seem to leave no doubt of its reality." This confident claim was based on the assertion of an officer stationed in northern India that "the unicorn exists . . . in the interior of Thibet, where it is well known to the inhabitants." They described it as "exactly the unicorn of the ancients."[195] A central African adventurer reported in 1862 that his skepticism had been "partly shaken" by the unanimous assurances of people living near the Niger River that they knew of a "strange, one-horned animal" and could show him its bones.[196] In 1873, a Natal correspondent of *Nature* relayed observations gathered from the Bushmen, concluding that "the unicorn existed recently in Africa, and that it is *not proved* to be extinct now."[197] From time to time, other sightings or near sightings of one-horned ungulates were reported from the Sudan and from Mozambique, among other places.[198] And even if unicorns did not occur in the normal course of events—that is, as a wild species—that did not necessarily reduce them to mere figments of the human imagination. Charles Gould suggested that they might occur in the abnormal course of events, as the fruit of a deer-horse hybridization. Alternatively, draping himself in the authority of taxonomic reification, he speculated "on the unicorn being a generic name for several distinct species of (probably) now extinct animals; missing links between the three families, the Equidae, Cervidae, and Bovidae."[199]

Mermaids and mermen boasted no more persuasive historical or scientific affidavits than did the elusive unicorn, but they were much more likely to present themselves to the British public in physical form. Perhaps for this reason, they were also more likely to undergo scientific analysis. Most zoologists were content to ignore reported unicorn sightings, which almost always, even when submitted by travelers with some credibility as field naturalists, came attended by an air of gossip, wishful thinking, and nostalgia. Exhibited mermaids, on the other hand, concretely challenged the established order of nature, which offered them no place. Such provocative public demonstrations predictably inspired at least commentary and evaluation, and often more forceful responses.

When one Captain Eades, the proprietor of a stuffed mermaid that fascinated London late in 1822, applied for expert anatomical advice, he did not apply in vain. Although he approached Everard Home before the creature had actually gone on public display, her reputation had prepared her welcome. The mermaid had been displayed at Capetown en route to Britain, and a British missionary had sent back an excited, detailed, and confused report, which had appeared in the metropolitan newspapers. It was, as William Clift later reflected, an illustration of "how easily persons may fall into error, and mislead others, when they attempt minute description of Subjects of which they have not a competent knowledge." Assuming that Home did possess such knowledge, Eades urgently requested his assessment of the genuineness of the mermaid, to pay for which he had sold the ship he commanded, and Home deputed Clift to perform the inspection.

Clift therefore met the captain one evening at the East India Baggage Warehouse. No sooner had the mermaid emerged from its carefully layered silk wrappings than he pronounced it a fraud, constructed of the cobbled remains of an orangutan, a baboon, and a salmon. Further examination revealed additional tampering, for example with the nails, which were synthetic, and the arms, which had been shortened. Nothing dismayed, Eades proceeded to install his treasure at the Turf coffeehouse, where it could be admired by anyone willing to part with a shilling, of whom there turned out to be several hundred each day at the height of the mermaid's popularity. Indeed, although his application to so exalted a source as Home for advice suggested sincere confidence that the creature was genuine, Eades's desire for an expert opinion may have been prompted less by reluctance to impose on the credulity of the British public than by the

A mermaid as analyzed by William Clift.

impending legal claim of the majority owner of the bartered ship, who wanted his fair share (which would have come to seven-eighths) of the gate. And in the course of the proceedings, the Court of Chancery did consider the question of authenticity, although the Lord Chancellor dismissed it in his decision, stating that "whether man, woman, or mermaid, if the right to the property was clearly made out, it was the duty of the court to protect it."[200]

For naturalists, however, the authenticity stakes were higher than they were for the Lord Chancellor. Home and Clift chose to treat Eades's request seriously, even though there was nothing unusually convincing about this "disgusting looking" and "frightful monster," as one journalist called it.[201] Indeed, the specimen did not even offer the lure of novelty. Among the many exotic and titillating curiosities that had competed for the attention of Londoners for the previous century and a half were a number of mermaids, most of them apparently manufactured in Japan for an Asian market.[202] On display within several decades of Eades's marvel, for example, were "The Wonderful and Surprising Mermaid, Angel Fish, Or Sea Woman! From Bengal" and a merman whose publicity cited many additional examples of "those extraordinary productions of nature" to "shew, that Mermaids and Mermen form a part of the creation."[203] At about the same time as Clift's devastating deconstruction of Eades's treasure, Thomas Hardwicke observed a similar creature in Calcutta, which, he matter-of-factly reported,

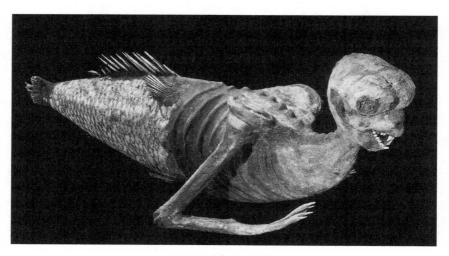

A material mermaid.

had "the head and arms of a Monkey—Teeth belonging to other Animals
. . . [and for the bottom an] unaltered part of a large Fish."[204] To such
concoctions one skeptic pedantically objected that they "would have required two distinct species of circulation, for a *warm*-blooded animal could
never coalesce with a *cold*-blooded one."[205] A more down-to-earth critic in
the *Annals of Sporting and Fancy Gazette* complained in 1824 that "another
mermaid has been brought to town from the Indian seas. Were she not dead
it is thought she could speak Sanscrit. Whereabout will the gullibility of
John Bull terminate?"[206]

Nevertheless, Home and Clift were not alone in considering such tawdry
displays worthy of authoritative scrutiny. Whatever their assumptions about
the likely existence of mermaids, zoologists were inevitably drawn to examine these multiply liminal creatures. In his *Curiosities of Natural History,*
Frank Buckland analyzed a mermaid he had seen at a Spitalfields public
house in 1858 as part simian and part hake. But even such dismissals had a
grudgingly respectful subtext.[207] More serious still were the attempts, which
appeared frequently in nineteenth-century natural history books, to account
for reported sightings of live mermaids (mostly by either sailors in exotic
seas or by the indigenous inhabitants of northern Scotland) and thus to
relocate potentially anomalous data inside the conventional confines of
natural history.[208] In 1799 the *Naturalist's Pocket Magazine* assured its readers
that the seal was "the true and sole foundation of the Mermaid"; in 1823
Robert Jameson, the Regius Professor of Natural History at Edinburgh,
similarly told fellow members of the Wernerian Society that the dugong
"may probably have given rise to the fable of the Indian mermaid"; and
Henry Lee, the naturalist in charge of the Brighton Aquarium, reminded
visitors to the Great International Fisheries Exhibition of 1882 that both
seals and manatees had provided "the originals of these 'travellers' tales.'"[209]
Indeed, the association of sirenians with mermaids became so conventional
that the conflation sometimes worked in the other direction. In this way,
experts could reappropriate mermaids in linguistic form, if not in their
physical incarnation. Thus *Notes and Queries* blandly reported that "the rib
of a mermaid is preserved in the vicar's library at Denchworth, Berks.";
under the rubric "Eating a Mermaid" readers discovered that the flesh of
the dugong tasted like veal.[210] As St. George Mivart observed at the end of
the Victorian era, "the dugong and the manatee are the only two mermaid
kinds now existing."[211]

Some naturalists, however, were willing to entertain the possibility that
the mermaids of legend, or something quite like them, actually existed. In
his *New Dictionary of Natural History,* William Frederic Martyn judged that

"there seems to be sufficient evidence to establish its reality." Several generations later, the quinarian taxonomists William MacLeay and William Swainson found a space in their artificial and elaborately embedded system of categories for an as-yet-unknown amphibious primate which might, as MacLeay suggested in 1829, "explain why there is such a general feeling among mankind of all ages, in favor of the existence of mermaids and may indeed render the past or present existence of amphibious primates probable."[212] Swainson more confidently asserted that "that natatorial type of the *Quadrumana* is most assuredly wanting. Whatever its precise construction . . . it would represent and correspond to the seals in the circle of the *Ferae* . . . that some such animal has really been created, we have not the shadow of a doubt."[213] And this receptiveness persisted among at least a few naturalists until quite late in the nineteenth century, although it was distinctly a minority point of view. The open-minded Charles Gould suggested that "many of the so-called mythical animals . . . come legitimately within the scope of plain matter-of-fact Natural History" on the grounds that both explorers and paleontologists were constantly discovering creatures that would previously have been implausible or even unthinkable.[214] As Gould's argument implied, mermaids seemed more spectacularly apocryphal when considered in isolation than when viewed in the context, or even in the company, of assorted other wonders, some of which, such as Siamese twins or platypuses from Australia, were incontestably real, if surprising. As the *Zoological Magazine* pointed out in 1833, even the early descriptions of the giraffe had seemed to combine such improbable combinations of characteristics that "within sixty years of the present time . . . its very existence [had been] cast into doubt."[215]

By the middle of the nineteenth century, stuffed mermaids no longer offered a compelling alternative to the established zoological order, and, as a consequence, serious discussion of them had become uncommon. That diminished attention did not, however, mean that the previous disorderly polyphony had been replaced by an authoritative and univocal source of information about the natural world. If mermaids had ceased to challenge scientific hegemony, other similarly mythological creatures rushed in to fill their places in Victorian hearts and minds. That is, there was a rather widespread willingness to accept evidence of other creatures that similarly violated the territorial limits set by zoology, and a similarly widespread inclination on the part of many zoologists to respond contentiously to these incursions into their domain. If the rapid exploration of the globe made it increasingly difficult to imagine a terrestrial habitat for creatures that defied the established laws of nature (the late-eighteenth-century discovery of the Australian fauna to the contrary notwithstanding), the ocean remained an unplumbed mystery. For most of the Victorian period, therefore, the liminal creatures of choice were the kraken and, especially, the sea serpent.

Newspapers and magazines readily offered space whenever shipboard or coastal observers sighted what Philip Gosse referred to as "the great unknown."[216] Convergent accounts, or an unusually spectacular or well-attested emergence, inspired particularly thick coverage, as was the case with the repeated sightings off the Massachusetts coast between Boston and Cape Ann in the 1810s, with the sighting by Captain Peter M'Quhae and the crew of the frigate *Daedalus* off the Cape of Good Hope in 1848, and with the sighting by Commander Hugh L. Pearson and the crew of the royal yacht *Osborne* near Sicily in 1877. But sightings were reported year in and year out, by fishermen, naval officers, and idle travelers. They spotted sea serpents at the four points of the compass and throughout the seven seas.[217] As *Nature* reported wearily in the late summer of 1872, "the 'dead season' has brought up its usual crop of reports of the re-appearance of the sea-serpent."[218]

If *Nature* was inclined to attribute most (but not all) of these reports to unusual accumulations of seaweed, journalistic discussion of sea serpents was often more respectful. The language in which sightings were recounted even in the popular press tended toward the sober and technical. The *Daily News* thus printed verbatim Captain M'Quhae's report to the Admiralty, complete with latitude and longitude, precise meteorological observations, and trigonometric estimates of the sea serpent's dimensions and speed.[219] In response to "the interest attached to this much vexed question," the *Illustrated London News* expansively contextualized the *Daedalus* sighting, so that the original report was supported by corroborative testimony from naturalists and naval officers who had seen similar creatures, summaries of New England sightings earlier in the century and of numerous Norwegian sightings in still remoter times, and detailed illustrations of M'Quhae's "enormous Serpent," as well as the apparently similar "Great American Sea-Serpent, *Scoliophis Atlanticus.*"[220] Since no sea serpent, or anything that persuasively resembled one, was ever captured, either dead or alive, they were never subjected to sideshow hype. And although the sensational potential of sea serpents—epitomized by a lurid *Police News* illustration of the sea serpent seen by the crew of the *Hydaspes* in 1876, in which the ship was dwarfed by a sinuous foregrounded monster that combined the most alarming aspects of mammals and reptiles (bristly hair, wild staring eyes, multiple jagged teeth, enormous size)[221]—was never entirely absent, most artists' renditions were much closer to the conventional, and relatively restrained, giant snake.

Despite persistent suspicions that sea serpents fitted more easily among creatures of oceanic myth than among quotidian marine animals, naturalists were also inclined to take them seriously. Charles Lyell's attention was captivated by the Boston-Gloucester sightings. For the next three decades he accumulated press cuttings and wrote letters in pursuit of information not only about the transatlantic serpent but also about such related creatures as "the (so-called) *Great Seasnake* of Stronsa," which had washed ashore in the Orkneys in 1808. A few caudal vertebrae had been preserved, on the basis of which some contemporary members of the Wernerian Society had assigned the creature the binomial "*Halsydrus Pontoppidani*" (after the eighteenth-century Norwegian bishop and pioneering sea serpent chronicler), but Lyell's more cautious correspondent admitted that he had "abandoned the hope of being able to establish either a generic or specific character, and therefore committed my notes to the fire."[222] But the fact that no identification had yet been made did not mean that none ever would be. Henry Lee suggested that although most sightings could be explained with reference to "known animals, especially if we admit . . . that some of

A lurid sea serpent.

them . . . may . . . attain to an extraordinary size," it would be "quite unwarrantable to assume that naturalists have perfect cognizance of every existing marine animal of large size."[223]

The apparent consensus represented by this zoologized discourse was, however, misleading in several respects. Like earlier discussions of mermaids, it conflated the views of people who believed that sea serpents were unknown animals that would prove, upon examination, to be similar to known animals (whether these were living or extinct, fish, reptile, or mammal) and the views of people who believed that sea serpents would prove, upon examination, to be dissimilar to known animals, and more like the creatures of legend. Further, although at least the former position—which can perhaps be characterized as cautiously open-minded—was widely held among both specialists and the interested general public, it was not the only discourse in which sea serpents figured. Throughout the nineteenth century reports of sea serpents triggered at least as much incredulity as respect. The likelihood of ridicule varied inversely with the esteem that the eyewitness commanded. The London press was particularly apt to greet American sightings with derision,[224] and members of the British lower classes were also automatically suspect. The fear of public ridicule as "grog-laden mariner[s] . . . exhibiting that phenomenon known to physiologists as 'unconscious cerebration'" almost certainly inspired many ordinary sailors to keep their sightings to themselves.[225] The evidence of British naval officers was, however, eagerly received. Sightings from what were perceived as reputable sources were taken so seriously that when the Commander-in-Chief of the *Osborne*, along with two of his lieutenants and the second engineer, submitted reports describing "a sea monster seen off the north coast of Sicily," the Admiralty passed them on to the Home Office, particularly requesting the opinion of Frank Buckland, the Inspector of Fisheries.[226]

Sober examination of such reports was not always viewed as the most reliable means of separating fact from fantasy. Some experts feared that it might instead act as the thin edge of the wedge, legitimizing the inclusion of the fabulous within the realms of science. Thus the tolerant skepticism or willingness to reappropriate misunderstood observations expressed by Lee or by Robert Hamilton, who judiciously commented that "till favouring circumstances bring the animal under the observation of Naturalists, the satisfaction which is desiderated respecting it, is scarcely to be expected," was countered by a rigid and principled rejection of all such data.[227] Many zoologists, among whom Richard Owen was the most prominent, viewed the persistent press coverage of sea serpents, and the widespread interest and credulity with which it was received, as a serious institutional challenge, and almost a personal affront. Owen was one of the experts consulted by the Home Office in 1877; his response, more dismissive than that of any of his colleagues, was simply to impugn the credibility of the witnesses.[228] But contemptuous as he might have been in public, he took the threat posed by the widespread willingness, and even desire, to believe in sea serpents very seriously. For more than forty years, from the 1830s to the 1870s, this extremely busy and powerful man took the trouble to keep a scrapbook in which he accumulated and annotated a variety of materials—newspaper and magazine cuttings, reports, letters, and memoranda—bearing on this issue. Owen's marginalia suggested the source of his antagonism, and that of many of his colleagues; one of the later entries asserted that "In proportion as persons are ignorant . . . and have their minds disturbed by ideas of monstrosity founded on phenomena which they don't comprehend [and] by emotions excited thereby, they are liable to confound facts [and] inferences [and] both to see, infer, [and] describe wrongly."[229] With its strong emphasis on background and methodology, this assertion discriminated sharply between those who rightfully occupied professional territory and rival claimants.[230]

The sea serpent, like the unicorn and especially the mermaid, embodied alternative modes of understanding the natural world. In a way, despite the fact that they received greatest attention at a time when the scientific establishment was more firmly entrenched, sea serpents offered a more plausible alternative than did mermaids. From the perspective of institutionalized science, they were more dangerous. After all, reports of mermaids, relatively small and elusive as they were believed to be, never counted as serious evidence; only material mermaids, such as the one examined by Clift, needed to be reckoned with, and these could invariably be debunked. Since sea serpents were never either captured or counterfeited, their pretensions could not be exposed to the standards of the laboratory. Indeed, there was no evidence of them at all, apart from the tradition composed of the accumulated reports of eyewitnesses and the assertions of experts based on those reports. Sea serpents existed in the realm of imagination and will and desire. These creatures of myth offered a blank text upon which people could inscribe their own beliefs about the organization of the animal kingdom, or, to look at it from a different angle, their reluctance to accept the structures imposed by self-constituted expertise.

Nor was this reluctance expressed only in negative or reactive ways. The hegemonic status of the formal classification of animals, whether cattle or kangaroos, crosses or monsters, was more obvious and more persusasive to its practitioners than to the uninitiated. Zoological taxonomy developed in eighteenth- and nineteenth-century Britain in continuing relation, whether parallel or antiphonal, to robust vernacular versions or alternatives. And some of these other modes of categorization, reflecting the most ancient and pragmatic relationships between people and other creatures, bore no relation to zoology at all.

5

Matters of Taste

Vᴇʀɴᴀᴄᴜʟᴀʀ ᴛᴀxᴏɴᴏᴍɪᴇs were numerous and diverse. Even within the realm of high culture, alternative expertise generated alternative classification. In elucidating the characters of both humans and animals, for example, artists proposed a range of doubles, some related tangentially to linkages grounded in zoological systematics, some ironically, some very little. The analogy between human beings and large felines exerted recurrent if various attraction. The late-eighteenth-century animal painter Sawrey Gilpin based his claim that "The lines, w[hich] form y. [the] countenance of y. lion approach nearer to those of y. human countenance, than y. lines of any animal with w[hich] we are acquainted," primarily on the animal's traditional role as the "King of y. beasts."[1] And although Gilpin arrived at this association by a classical or literary road, zoological techniques, severed from their institutional context, could produce similarly heterodox conjunctions. The romantic artist Benjamin Haydon derived his identical insight from anatomical exploration: "while dissecting a lion, . . . I was amazingly impressed with its similarity as well as its difference in muscular and bone construction to the human figure. It was evident that the lion was but a modification of the human being."[2] And if the moral similarity of people to lions or to chickens was primarily of abstract or philosophical interest, other taxonomic associations had more pragmatic implications for the conduct and understanding of social life. Among the most evocative and revealing were those that structured the killing and eating of animals.

Conflicts of Interest

British hunters sorted wild animals into elaborately overlapping sets of human-oriented categories. The most ancient divisions, which survived into the nineteenth century only slightly modified from their medieval origins, classified game according to the kind and degree of amusement it offered. For example, a Victorian sporting compendium cited Edward II's huntsman as authority for a division into "beasts for hunting," including the hare, the hart, the wolf, and the wild boar; "beasts of the chase," including the buck, the doe, the fox, the marten, and the roe; and "beasts that afford greate dysporte," including the badger, the wild cat, and the otter—despite the fact that only a few of these animals remained legitimate, or even possible, objects of British venery.[3] Sportsmen averse to anachronism could easily modify the categories in accordance with current availability and conventions; or they could generalize them to accommodate game undreamed of by hunting knights and squires. In a handbook that grandly included a "synoptical guide to the hunting grounds of the world," the taxidermist Roland Ward divided game animals into "two classes: (a) those that are dangerous, (b) those that are not seriously dangerous," within both of which could be distinguished "(c) animals that are in natural condition unsuspicious, or quiescent; [and] (d) animals infuriate, aggressive, charging."[4] Categories that differentiated game species according to age and sex, in parallel to categories available for most domestic livestock, expressed a different sense of the hunter-prey relationship, one that emphasized not violence and antagonism, but stewardship and possession. As "the first year we call a ewe a lamb; the second year a ewe pug or teg; the third year a thaive; and the fourth year a sheep" and the "wether we call the second year a wether pug or teg; the third year a sherrug, and the fourth a sheep"; so the stag "is called, in the first year, a *calf,* or *hind-calf;* the second, a *knobbler;* the third, a *brock;* the fourth, a *staggard;* the fifth, a *stag;* and the sixth, a *hart.* The female is called, the first year, a *calf;* the second, a *hearse;* and the third, a *hind.*"[5]

If such classifications emphasized the varying connections between human predators and their prey, hunters also assigned animals to categories that referred explicitly to human relationships. The notion of "game" itself, fundamental to all kinds of hunting, defined not only which animals could legitimately be taken but which people could legitimately pursue them. Originally connected with social position, after the passage of the Game Act of 1671 the right to hunt was firmly vested in land ownership.[6] Proprietors routinely retained this right when letting their land for agricultural purposes; increasingly after the middle of the eighteenth century, they further attempted to ensure that game would be abundant when it was wanted by establishing protected preserves. Both these developments identified the interests of landowners with those of the preserved animals (at least until the ultimate moment of truth) and against those of other human occupants of the countryside. Tenant farmers repeatedly lamented the depredations that protected animals made on their crops. In some areas preservation was carried so far that turnips and peas had to be defended against "hares . . . grey with age, with teeth turned up outside the jaw like the tusk of a wild boar."[7]

In addition the Game Laws, and the institutions that evolved to enforce them, offered two parallel taxonomies that were purely human. One was a hierarchy of venatic privilege, capped by those rich enough to own land. A mid-nineteenth-century critic in the *Westminster Review* thus castigated the Game Laws as "class-legislation," enacted "simply that at certain seasons of the year their self-constituted owners may enjoy the dignified pastime of slaughtering by hundreds a parcel of defenceless and semi-domesticated creatures,—a species of amusement from all participation in which the *profanum vulgus* are rigidly excluded."[8] As legitimate access to game categorized people on a scale of social prestige, illegitimate access categorized them on a reverse scale of criminality. Depriving ordinary country dwellers of the legal right to kill even hares and pheasants, which in many areas were both abundant and destructive, redefined more or less law-abiding populations as habitual transgressors.

By the early nineteenth century it was "notorious" that "among the poorer classes in the game-preserving districts the crime of poaching is almost universal."[9] But if lower-class poachers were the most aggressive, disruptive, and severely punished violators of the Game Laws, many of their betters were guilty of subtler infractions. The increasing prevalence of poaching after the middle of the eighteenth century, and of the violence associated with it, reflected more than political resentment against landlords or the quest for dietary supplementation.[10] Poaching offered significant financial rewards, because it supplied the tables of largely urban consumers rich enough to pay a premium for prohibited game, even if they lacked the funds or the inclination to purchase the land on which it lived. Although the sale of game was illegal, it was readily available through ordinary commercial channels. In 1824 one London poulterer claimed that he could "provide every family in London with a dish of game," and another that "he would engage to supply the whole House of Commons, without the least difficulty, twice a week for the whole season."[11] Heavily armed midnight trappers were only the first link in a network of distribution and marketing that criminalized every level of rural society. Many gentlemen who prosecuted poachers in the morning fenced stolen goods (even if the goods had been, in a sense, stolen from themselves) in the afternoon.

Hunting was an aggressively expansive activity, one that abutted on or intruded into many other aspects of bucolic life. Even the simplest of hunting categorizations—the dichotomy between game animals, which had legal protection and were appropriate objects of the chase, and nongame animals—begged questions of demarcation and value that people with other interests in animals and the countryside answered very differently. The consequences of these disagreements extended beyond opinion and attitude, into the realm of life and death. Each set of categories dictated its own boundary between protected animals and animals whose elimination was licensed or required.

If the benefits of the elevated status of "game" were questionable from the animals' point of view, no advantage necessarily accrued from exclusion. Although most animals that were not legally game were not targeted by hunters, neither were they defended by them. Anyone could kill them, in any way, for any reason. Often, the alternative to being labeled as "game" was to fall within an equally stringent dichotomy, based on utility rather than diversion. Creatures that, far from being helpful or profitable, seemed actively pernicious were stigmatized as "vermin" and doomed to extermination. An early-nineteenth-century naturalist thus recommended that "on procuring an Animal with which we are unacquainted, the first point . . . is to ascertain whether it is . . . applicable to the uses of Man . . . , or should its habits be detrimental or obnoxious, what measures are pursued to destroy the species."[12] Behind this recommendation lay the assumption, encouraged by both theology and economics, that the world had been created for the benefit of humankind, and that it was the prerogative, even the duty, of people to remedy apparent lapses in these arrangements. They only had to be careful not to jump to hasty conclusions. According to a mid-Victorian professor of natural history at the Royal Agricultural College, conscious of the vigor with which farmers were apt to fulfill this obligation, only some anathematized animals were exterminated "upon good evidence of felonious deeds, whilst others are the victims of prejudice or mistaken observation."[13] A contemporary professor of agriculture similarly cautioned that "it must not be supposed that [animals] are useless productions, because an abundance of them proves injurious to the specific object of the farmer," and that even if "no immediate application emerged" a creature "may answer some purpose that has escaped our observation." Nevertheless, he concluded that "it seems reasonable to destroy . . . the least useful."[14]

One reason that taxonomic errors were frequent was that the category of "vermin," although potent in its implications, resisted clear definition or demarcation; on the contrary, it "comprehended a wide range of creatures, having little affinity with each other."[15] Most authorities failed to offer any general characterization that referred to the nature of the animals, as opposed to their effect on human property; they preferred instead simply to list the offenders. Such catalogs tended to be comprehensive. An early-eighteenth-century agricultural manual itemized not only the pernicious animals but also the depredations in which they specialized: foxes and badgers targeted lambs and poultry, otters fish, hares corn, polecats pigeons, and rodents all crops. In 1768 the self-styled "rat-catcher to the Princess Amelia," who hoped that his treatise on vermin would "be productive of great public utility although the subject is but low and humble," specified the fox, the otter, the badger, the sheep-killing dog, the wild cat, the weasel, the ferret, the stoat, the marten, the hedgehog, the bat, and a variety of rodents, as well as hawks and owls, the crow, and the magpie.[16] The list offered by a Victorian naturalist was more or less identical, although the vernacular names were supplemented by latinate binomials; it included almost every wild mammal native to the woods and fields of Britain, and many of the birds.[17]

The taxonomy of farmers projected a placid, orderly, controlled country-side free from animal interference and competition, while the taxonomy of hunters projected a countryside full of surprises, excitement, and even danger. Since they targeted different creatures, however, in practice game-based categories and vermin-based categories often coexisted peacefully, even within the mind of a single individual. But one man's vermin was occasionally another man's game. The lagomorphs and birds protected by the Game Laws were thus "destructive" and "hurtful" from an agricultural perspective, and the sport derived from slaughtering them condemned as "so contemptible as hardly to deserve notice."[18] Indeed, it was alleged that "the very mention of hares and partridges . . . too often puts an end to common humanity and common sense."[19] Although foxes were not legally defined as game—they were, on the contrary, incontestable vermin—they were similarly cherished as occasions of sport. In areas where the landlords rode enthusiastically to hounds, farmers were constrained by self-interest if not by law to absorb repeated losses of poultry and young animals, as well as the destruction to fields and gardens caused by the unpredictable gallopings of the hunt itself. Assessments of the costs and benefits of foxhunting, like those of game preserving, strongly reflected the categories, and therefore the values, uppermost in the minds of commentators. From a sporting perspective, although foxes were "great pests . . . yet the excitement of an English foxhunt, especially to the quiet and retired agriculturist, is frequently allowed to be a sufficient set-off"; but to the pragmatic farmer, it was "a relic of a barbarous age, when the uncultivated state of man 'ranked him among the ferae naturae.'"[20]

However much they disagreed about foxes, hares, and pheasants, farmers and hunters agreed in condemning another group of animals. Authors of natural history manuals tended to vilify the carnivores, characterizing them as generally "injurious" or plainly stating that they "rank as vermin."[21] But wild foxes (and badgers, wild cats, weasels, polecats, and stoats) were not the only carnivores wandering loose in British fields and farmyards. They were joined by pet dogs and, especially, cats, many of whom shared their predilection both for poaching game and for stealing chickens.[22] From the point of view of hunters and farmers, this habit made them vermin, distinguished from other noxious animals only by their capacity to "do infinitely more mischief than many vermin naturally wild."[23] As a result, they frequently met the fate of vermin, trapped, shot, or poisoned. They figured prominently in the "museums" assembled by vigilant gamekeepers, who nailed the skins of slain predators to the sides of barns. Frank Buckland counted the heads of 53 cats in a single collection, noting with interest that each displayed a "different expression of countenance."[24]

Labeling these unfortunate animals as vermin, however, begged an important taxonomic question. In the eyes of their owners they were classed as pets, a category that reflected yet another kind of relationship between humans and animals, based on the dichotomy between the domestic and the nondomestic. Neither the category nor the relationship was of great interest to farmers and hunters, who tended to think of their own animal companions, such as sheepdogs or hounds, as fellow laborers or willing servants, and to disparage a merely affectionate connection with creatures they regarded as idle and useless.[25] When they privileged their own taxonomies and destroyed errant cats and dogs as vermin, not only did they violate bonds perceived by pet owners as nearly familial, but they rejected the subtly suburbanized countryside, dotted with cozy but unproductive households, that the category of "pets" also implied. In a *Westminster Review* article ironically titled "Unprotected Vermin," the bereaved owner of a Persian cat that had been trapped and bludgeoned to death by her game-preserving neighbor insisted on the "pecuniary value" of the animal, as well as the right of pet owners to consideration equal to that received by representatives of other rural interests. She noted that "just as . . . for the fox," it was "perfectly natural" for the cat to "occasionally help himself to a little game," and suggested that those whose yards were raided should remember that "we, that is those who neither hunt nor shoot, have to tolerate the depredations of our neighbours' game." But her strongest protest insisted on the legitimacy and value of this category of animals, emphasizing qualities peculiarly appealing in the context of pet ownership, although meaningless in the context of sport or utility. The lost animal was "magnificent," "exquisite," "beautiful," and "defenceless"; it was "loved and petted, and . . . well-nigh indispensable."[26]

You Are What You Eat

Natural history and comparative anatomy, hunting and pet keeping, engaged only segments of the British population—albeit, in the case of at least some of these activities, substantial segments—but participation in the production, preparation, and consumption of animals as food was nearly universal. As in many cultures, the significance of meat extended far beyond the efficient provision of calories.[27] In the same way that "Carnivorous Animals have more Courage, Muscular Strength and Activity," so a meat-heavy diet was believed to contribute to British "energy, . . . sense of fitness, . . . [and] craving for work . . . The beef of the British soldier has always been regarded as a factor in his valour."[28] Meat eating was viewed by observers within the British polity and outside it as an essential component of the national character. Well into the second half of the nineteenth century, Britain's "foreign neighbours" were reported to "believe that all classes in England, excepting 'La grande aristocratie,' live upon nothing but half raw roast beef and steaks, plum pudding, and brandy 'grogs.'"[29]

Visitors from Europe also noted with astonishment the amount of flesh regularly eaten in those British households able to afford it.[30] Abundant meat was considered not a luxury of domestic economy, as it might be in less fortunate or carnivorous lands, but a necessity. The sample weekly budgets included in early-nineteenth-century editions of Mrs. Rundell's System of Practical Domestic Economy assumed that even families of rather modest means would provide each member with more than half a pound of meat each day; several decades later a cookbook writer estimated that every resident of London consumed 107 pounds of butcher's meat per year, far outstripping the 85 pounds consumed by the average Parisian.[31] When asked by her mistress what she and "her father and brothers ate as a rule

Veal

Beef

Mutton

Pork

Turtle

Venison

Gastronomic taxonomy.

for dinner," a Victorian maid replied, "'Why beef and mutton, ma'am, every day! Working people *want* their meat."[32] William Cobbett warned of the dire consequences of its absence in the poorest households, claiming that "*Meat in the House* is a great source of *harmony,* a great preventer of the temptation to commit those things, which, from small beginnings, lead, finally, to the most fatal and atrocious results."[33]

Cobbett saw salvation for the needy mainly in terms of bacon, "a coarse and heavy, but nutritive food," which, in the view of many dietary experts, was "only fit to be taken in considerable quantities by robust labouring people."[34] The iconic meat of the British, however, was beef. As the *Quarterly Journal of Agriculture* complacently noted in 1830, "our domestic cattle . . . [are] the most varied and remarkable in the world, and have long yielded us . . . good beef (of which the very name is almost identified with the character and propensities of the nation)."[35] According to *Household Words,* "beef is a great connecting link and bond of better feeling between the great classes of the commonwealth," inspiring respect second only to "the Habeas Corpus and the Freedom of the Press."[36] As it symbolized the common predilections and loyalties of Britons, so beef emphasized the irreducible alterity of foreign cultures. Thus, "on the Continent . . . tasteless pieces of beef which have been used to make the bouillon are invariably served with sauces . . . , and the *bifsteak,* so called in honour of the famous English dish, is often a piece of very coarse buffalo"; while "John Bull in India . . . sighs for a Southdown saddle [of mutton] or a Scotch sirloin, and is apt to turn away sorrowfully from the meagre travesty of a joint which, after much trouble, the sharer of his joys and sorrows contrives to place before him."[37]

Naturalists and agricultural experts, chefs and gourmets agreed that Britain was preeminent for the quality of the meat it produced, as well as the quantity it consumed. This additional superiority confirmed the binary distinction between the British and peoples apparently less cathected to animal flesh. Discussions of the national livestock inevitably reiterated a conventional litany of celebration. This consensus asserted that British cattle were "considered preferable to the cattle of any other country in the world"; that "the cattle and sheep of this country we may justly regard . . . as unequalled in any other territory"; and that "no country produces finer sheep than Great Britain."[38] Even a cookbook whose irreverent author disparaged the finished products of British cuisine acknowledged that its raw materials were unrivaled: "In this country we have the best of all descriptions of butcher's meat in the world, and, with a few exceptions, the worst cooks. If the poor, half-fed meats of France, were dressed as our cooks, for the most part, dress our well-fed excellent meats, they would be absolutely uneatable."[39]

This patriotic consensus was not, however, universal. The eighteenth century, when the association of Britain and beef was consolidated, also saw the first stirrings of the modern vegetarian movement. Vegetarians argued either that meat eating did not lead to the desirable consequences, whether personal or national, with which it was associated or, more radically, that those consequences were not desirable after all. According to a mid-eighteenth-century account of the "Pythagorean diet" of fruits, vegetables, milk, and honey (and, in a relaxation of classical strictures, "some very moderate Portion of animal Food . . . provided it were . . . young and tender"), the fleshy fare prized by most Britons was unhealthful and worse, leading to such "bad and venomous Consequences" as "Wars, Sieges . . . , long Encampments, [and] . . . remarkable Pestilences."[40] Half a century later John Frank Newton warned that "we Englishmen, who rival all nations in attachment to solid food, are remarkably subject to perish by contagion in hot climates."[41] In some observers, the production of meat for the table evoked not just valetudinarian concern but active repulsion. At the beginning of the nineteenth century a drawing manual that featured scenes from everyday life omitted the butcher as "no very fit subject for the artist . . . There is something in the profession . . . exceedingly shocking."[42] The vegetarian crusader Joseph Ritson contended that the "sight of animal food is unnatural and disgusting," inevitably a reminder of the dead body from which it had come.[43] A Victorian pamphlet warned meat eaters that "scientific experiment has . . . shewn that . . . no portion of animal flesh . . . is free from excremental matter which at the moment of killing is arrested on its way to discharge from the system."[44]

Persuaded by such emotional appeals or by more carefully reasoned arguments—which sometimes included comparisons of human teeth, stomachs, and other organs with those of animals known to be frugivorous or vegetarian, such as monkeys and apes—the ranks of vegetarians grew slowly through the nineteenth century.[45] By the late Victorian period, they were sufficiently numerous to be organized into societies, to provide a discernible sub-market for the booming cookbook industry, and to patronize restaurants with menus wholly or partly devoted to their preferences.[46] Catering arrangements suggested that the appeal of vegetarianism was felt throughout the social hierarchy. Some London "people's cafes" served cheap meatless dishes: plain vegetables for 3 pence, vegetable soup for 4 pence, and such savories as braised vegetables, stewed lentils, and potatoes maître d'hôtel for 6 pence.[47] Charles W. Forward drew on the recipes of more pretentious establishments, with names like the "Porridge Bowl" and the "Apple Tree," to construct the menu for an elaborate formal dinner, in which the entrées included tomato toast, mushrooms en papillote, celery compote, and rissolettes of potatoes.[48] Three thousand members of the Vegetarian Society attended its conference in 1881.[49] During the period that vegetarianism was thus evincing its attractiveness, the amount of meat routinely consumed by carnivores declined somewhat from its plethoric peak.[50] But mass voluntary abstinence never threatened the beef-oriented norm. At the end of the nineteenth century, Forward could only wonder disconsolately why, since "the humane diet has so many powerful arguments in its favour . . . the persistent advocacy and example of vegetarians have not achieved larger results."[51]

Whether or not Forward fondly overestimated the charms of vegetarianism, the symbolic associations of meat eating may well have exacerbated routine dietary conservatism. Rejection of the national diet smacked of profounder dissents; if beef eaters were self-classified as British, abstainers might belong in more disturbing categories. Even modest vegetarian suc-

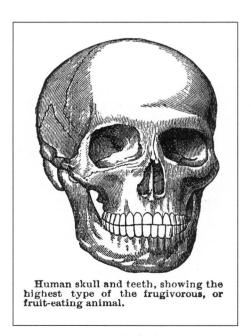

Human skull and teeth, showing the highest type of the frugivorous, or fruit-eating animal.

Anatomy is destiny.

cesses, or vegetarian suggestions unlikely to exert any noticeable influence on public behavior, could engender patriotic unease, and more forceful and systematic exhortations could trigger patriotic fury. Although some vegetarian advocates argued in reasoned tones for limited health-oriented ends, others stridently challenged not only meat eating but the entire social and political context that it represented. For example, a rather sympathetic account of Joseph Ritson's campaign against meat noted that he "persevered in [it] with the same inflexible determination which guided all his actions," and that it exposed him to a certain amount of contumely and ridicule.[52] A hostile notice of Ritson's *Essay on Abstinence from Animal Food, as a Moral Duty* in the *Edinburgh Review* more clearly suggested the intensity and the sources of such criticism. The anonymous author reported himself stirred by "disgust, pity, contempt, laughter, detestation" as he read the tract, and he proposed to share with his audience "the various emotions which it has raised in our minds," rather than its arguments. As a habitual carnivore he indignantly rejected Ritson's critique of his order, protesting ironically that his diet had not "diminished that gentleness and placidity of temper that belongs to our profession," he objected with real outrage that "'*Eater of beef and mutton*,' is here used as synonymous with cannibal."

At the end of the nineteenth century Henry Salt embraced vegetarianism as part of a constellation of advanced social and political views, but it was the one which struck his Eton colleagues as most scandalous. He, in turn, was increasingly convinced that they were "but cannibals in cap and gown."[53] And to some devout believers, committed herbivory seemed to challenge the foundations of the church as well as the state. Orthodox eaters found it necessary to rebut the attempts of "various speculative authors" to give vegetarianism a scriptural basis by attempting "to prove that animal food was not eaten before the deluge, but was introduced in consequence of the deterioration which the herbage sustained on that occasion."[54] A fundamentalist critic alleged that vegetarian apologetics not only misinterpreted Christian doctrine but were "deeply and openly opposed to Christianity," obliging adherents to "resist God's command in the Lord's Supper . . . because the *wine represents blood*."[55]

The time-tested benefits of meat eating provided a more restrained riposte to vegetarian claims. It was widely acknowledged that, as a Victorian cookbook for convalescents put it, "animal food satisfied hunger more completely and for a longer time than vegetables," and that beef and mutton were the most nutritious of meats. Such endorsements often carried some suggestion of moderation, a recognition that, taken in excessive quantity these "noble viands . . . of heroic proportions" offered denser sustenance than "our modern civilised life" required, even as led by Royal Beefeaters.[56] Or they warned of the intellectual consequences of overindulgence: "whoever would keep his mind acute and penetrating, should rather exceed on the side of vegetable food."[57] But the possibilities, however remote and manageable, of indigestion and stupefaction were not universally admitted. The most enthusiastic meat advocates claimed that it was more easily digestable than lighter foods. The late Victorian patients who followed the regimen of the American physician James Salisbury regained their strength on a diet that, in its strictest form, included only egg whites, and black tea or coffee, and the eponymous Salisbury steak. As they recovered they were allowed to relax their regimen somewhat, adding chicken or lamb or even rice, but "only [as] a relish . . . the steak to be four-fifths of the meal." Patients were assured that once they embraced this diet, they would begin to crave wholesome meat as they had once craved unwholesome vegetables.[58] And some people attributed to carnivory still more surprising efficacy. For example, in 1860 the *Lancet* suggested that meat eaters had lighter, clearer complexions and that, therefore, "under a well-devised animal diet it might be possible . . . to whiten the blackamoor by feeding, though not by washing."[59]

If meat eating in general distinguished Britons from foreigners, it also provided occasions for reinforcing intranational categorizations. The superior sophistication of the metropolis, for example, distinguished its butchers from their rustic colleagues, who lacked "the secret of the proper cut."[60] Even in its unsegmented state, meat reflected essential social divisions, some elevated kinds "fit for fine Complexions, idle Citizens," others, like bacon, "gross, tough and hard, agreeing chiefly to Country Persons and Hard Labourers."[61] The hierarchy of class privilege expressed through access to furred and feathered game was echoed in cookbooks. Sometimes elite status was said to confer special nutritional qualities, so that, for example, venison, grouse, pheasant, and, indeed, "all game" were credited with a "looseness of texture" that made them "more digestible" than the flesh of domesticated birds and mammals.[62] Or the dichotomy was embodied in recipes, with venison, hares, rabbits, and game birds grouped for culinary treatment that emphasized their shared distinction from domesticated birds and livestock.[63] So strong were the associations between social prestige and game, and between game and its conventional preparations, that people who could not aspire to the real thing sometimes borrowed the recipes, metaleptically serving marinated beef in the guise of venison.[64]

More widely available meats were ranked not only by species but also according to other categories significant in a human context. When it came to ingestion, standard hierarchies of age and sex were inverted. The flesh of mature animals was coarser and tougher than that of delicate young ones, and their fat was less evenly distributed through their muscles. In addition, they had spent more time under what one physician described as "the very extraordinary . . . influence of the genital organs." Although he believed that even the ovaries impaired "the flavor of the female," the testicles were the main culprits: "every day the testes are permitted to remain . . . injures the delicacy of the veal of the bull-calf."[65] Experts in domestic economy generally recommended that "heifer and ox beef are both excellent" and that both were preferable to cow beef, as "a leg of wether mutton is more desirable than a ewe one."[66] One Victorian commentator forcefully objected to butchering "the females of . . . domesticated mammalia" on "the joint grounds of impropriety and unwholesomeness," but the meat of cows nevertheless ranked considerably higher than that of bulls, since "females are sweeter, moister, and easier to be concocted."[67]

In general, the meat of mature male animals was abominated. Bull beef was easily recognized for its "strong, unpleasant smell"; and "ram mutton

Original caption:
A GENTLE VEGETARIAN.
"'MORNING, MISS! WHO'D EVER THINK, LOOKING AT US TWO, THAT YOU DEVOURED BULLOCKS AND SHEEP, AND *I* NEVER TOOK ANYTHING BUT RICE?"

has a strong, and . . . exceedingly disagreeable flavour."[68] Hearsay evidence suggested that the opposition of delectable females and unpalatable males similarly occurred in some rarely eaten wild creatures; thus, according to Thomas Bewick, among "Sea-Bears" (a kind of seal) "the fat and flesh of the old males are very nauseous; but those of the females and the young . . . are said to be as good as . . . a sucking Pig."[69] A mid-Victorian agricultural encyclopedia attributed the "sanguineous odour and flavour" of rams and bulls to masculine behavior as well as masculine essence, being "aggravated by . . . the exercise of the procreative faculties."[70] Among livestock, only pigs—the occupants of liminal positions in many taxonomies—broke the gender pattern: "the flesh of the sow is strong," while "the flesh of the male . . . [makes] the best pork."[71]

The terms routinely employed to characterize masculine and feminine flesh, and to differentiate males in full exercise of their powers from their neutered brothers, recalled the stereotypical language associated with men and women. The aversion to viewing as mere table fare dominant bulls and rams—the leaders of herds and the fathers of families—may have derived as much from political as from culinary predilections. After all, the strong flavor that made male beef and mutton inappropriate for quotidian consumption seemed not only tolerable but desirable when eating was charged with the special symbolism of preeminence or domination. Thus game, emblematic of class privilege, was valued for exactly those qualities that disqualified the flesh of domestic male animals. Hare was superior to rabbit, "being much more savoury, and of a higher flavour," and "buck venison is preferred [to that of the doe] as the choicest meat."[72] As game was cooked in special ways, so it was prepared distinctively by butchers, to enhance its characteristic rankness. According to a plain-spoken Victorian cookbook, aficionados of venison were divided between those who "like it a little gone, and others a good deal. This state of putrescency is called by gourmands *haut gout*, high tasted; we should rather say at once, stinking."[73] Where eating was concerned, the allegedly wild white cattle of Scotland and northern England again bridged the categories of wild and domestic.[74] Despite their resemblance to ordinary farmyard beasts, their beef tended to be served only on ceremonial occasions, after a ritualized hunt. The Earl of Tankerville slaughtered a number of Chillingham cattle to celebrate his son's birthday in 1756, and the Victorian owners of the Lyme Park, Cheshire herd shot one or two animals each Christmas and forwarded some of the beef to Queen Victoria.[75] Although the flesh of these animals—"dark in colour and very full flavoured"—shared the attributes that made the beef of ordinary bulls unacceptable, it was nevertheless considered "excellent."[76]

You Aren't What You Don't Eat

If the consumption of some kinds of animals was mandatory for Britons, and the consumption of others lent social cachet, many creatures existed beyond the gustatory pale. Their proscribed number had tended to increase since the less finicky medieval period, when any wild animal or bird could end up in a pot or a pan.[77] Fifteenth-century cookbooks included recipes for whale, seal, crane, and peacock, and the taste for red squirrels among the aristocracy declined only in the Tudor period, when suspicions emerged that the rodents were carrying disease.[78] The flesh of speared otters was not only eaten but eaten during Lent and on other "maigre days, from its supposed resemblance to fish, on which otters almost wholly subsist."[79] The Naturalist's Library volume on cetaceans recorded the opinion of "Dr. Caius, the celebrated founder of the College at Cambridge, . . . that a Dolphin . . . was . . . a worthy present to the Duke of Norfolk, who . . . [had it] roasted and eaten with porpoise sauce," as an example of "the change of tastes produced by modern refinement."[80] Those who praised the heightened fastidiousness as progress associated it with social and religious order, as well as more delicate sensibilities. As John Goodsir approvingly instructed his comparative anatomy students at the University of Edinburgh, it was appropriate for civilized people to prefer foods "which were created for [their] special use—e.g. ox, sheep, goat, wheat, [and] barley."[81]

But what complacent consensus perceived as increasing refinement could be reinterpreted by dissenters as forgone opportunity or even waste. The circumscribed animal diet of eighteenth- and nineteenth-century Britain, conducive though it might have been to developing solid national character, was open to criticism on other than vegetarian grounds. Pragmatic carnivores reductively lamented the economic loss resulting from the neglect of readily available sources of nourishment. An eighteenth-century dietary reformer reasoned, "When I considered how cleanly the Hedge-hog feedeth, namely upon Cow's Milk, . . . or upon Fruit and Malt; I saw no reason to discontinue this Meat any longer upon some fantastical Dislike."[82] In 1885, V. M. Holt similarly began his tract *Why Not Eat Insects?* with the rhetoric of rational persuasion, asking his readers to substitute "a fair hearing, . . . an impartial consideration . . . , and an unbiassed judgment" for "long-existing and deep-rooted public prejudice." He assured them that the insects he recommended for consumption (in which category he included "small mollusks and crustaceans . . . for the sake of brevity and convenience") were all "clean, palatable, wholesome, and decidedly more particular in their feeding than ourselves."[83] And, lest reluctance to partake stem merely from ignorance of appropriate preparation, Holt offered several menus that featured insects in every course. *Potage aux limaces à la chinoise* (slug soup) was followed by *larves de guêpes frites au rayon* (wasp grubs fried in the comb), *phalènes à l'hottentot* (moths sautéed in butter), and *petites carottes, sauce blanche aux rougets* (new carrots with wireworm sauce), with *crème de groseilles aux nemates* (gooseberry cream with sawflies) for dessert, and *larves de hanneton grillées* (deviled chafer grubs) and *cerfs volants à la gru gru* (stag beetle larvae on toast) as savories.[84]

As Holt pointed out, the national aversion to unfamiliar food was not complete or consistent. During the period when he was composing *Why Not Eat Insects?*, he complained, a health exhibition in the metropolis featured a Chinese restaurant where patrons had happily eaten soup made from "the nest of a small swallow, constructed . . . principally by the means of . . . a viscid fluid secreted from its mouth. Does not that sound nasty enough?"[85] Nevertheless, despite the logic of his arguments, his ideas had little impact on the diet of his fellow citizens. And more sustained, better orchestrated, and less lonely campaigns produced similarly unimpressive results, which suggested that British neglect of readily available protein sources resulted from profounder causes than simple inattention or failure of nutritional acumen. The most systematic Victorian attempt to expand the category of edible animals targeted horses and donkeys. Advocates of hippophagy and onophagy, briefly united in the Society for the Propagation of Horse Flesh as an Article of Food, argued not only that a great deal of meat was being wasted—since dogs and cats could not consume the remains of all the horses who died after lives of toil in British streets, fields, and mines—but also that animals destined for the butchers rather than the knackers would experience less brutal treatment in the final months of their lives.[86]

Even in the cause of kindness, however, the British could not be persuaded to eat horses. Their resistance was acknowledged to have deep, although far from primeval, historical roots. Because the pre-Christian inhabitants of Europe had consumed horsemeat with ritual enthuasiasm, early evangelists banned it as part of the comprehensive suppression of pagan observances.[87] And if the practice had nevertheless intermittently reemerged in less prosperous provinces, it had therefore acquired an additional set of undesirable associations with scarcity and backwardness; in 1617, for example, it was reported with disgust that the "wild Irish . . . will feed upon horses dying of themselves, not only upon small want of flesh, but even for pleasure."[88] Rationalistic horseflesh advocates assumed that in the impregnably Christian societies of nineteenth-century Europe, the danger of recrudescent paganism would pale before such palpable utilitarian advantages as cheap protein and high resistance (compared with beef, for example) to tubercular infection.[89] In France and Belgium the campaign succeeded. The first horsemeat butcher shop opened in Paris soon after the gala *banquet hippophagique* held at the Grand Hotel in 1865.[90] But neither the francophone example, nor the similar practices of other European nations, such as Germany, Russia, and Denmark, where horseflesh reportedly constituted a "principle item" in the diets of prisoners, softened the intransigence of British diners.[91]

On the contrary, the foreign associations evoked by the horsemeat campaign were more likely to confirm insular resistance. The banquets organized in England in 1868 were multiply exotic. Not only did they feature a generally reprehended meat, but they were cooked in the French style by French chefs. At the heavily publicized dinner sponsored by the Society for the Propagation of Horse Flesh, the strangeness of the main attraction was exacerbated by the elaborate menu, in which the inevitably Gallic nomenclature was further laden with literary and zoological allusions. The *potages* included *purée de destrier;* the *poissons, saumon à la sauce arabe;* the *hors d'oeuvres, terrines de foie maigres chevalines;* the *relevés, filet de Pégase;* the *entrées, petits pâtés à la Moëlle Bucéphale* and *poulets garnis à l'hippogriffe;* the *rôtis, mayonnaises de homard à l'huile de Rossinante;* and the *entremets, gateau vétérinaire à la Ducroix.* Even on less gala occasions, advocates gallicized the newly edible horsemeat as "chevaline."[92] And in any case the horsemeat campaign did not aim to change the eating habits of people accustomed to French menus and public dinners, but those of their less privileged, less urbane, and presumably less adventurous compatriots.

Nor did willingness to try chevaline necessarily ensure conversion. When the supposed leaders of fashion, or at least the supposed molders of the opinions and habits of their social inferiors, good-naturedly agreed to try horsemeat, their subsequent enthusiasm was seldom striking. Even dedicated advocates tended to express their appreciation in very moderate terms. The guests at a French chevaline banquet were reported merely to have been "perfectly satisfied," and the horseflesh on which Danish prisoners throve tasted "like coarse beef, yet by no means unpleasantly," when served plain, and was "although far from constituting an agreeable product, . . . not unpalatable" in soup.[93] Those less committed coded latent aversion as simple distaste. In 1887 a veteran gastronome admitted that, although he had attended more that one hippophagic banquet, "the impression made . . . on my palate and left on my mind . . . [was] not . . . pleasing." His dislike of chevaline was based on less concrete considerations than its flavor;

on the contrary, he usually "found the soup and other dishes tasty." But before beginning to eat he had to "force down my instinctive reluctance to turn the noble horse into an article of food," an effort that invariably proved too taxing for long continuation, so that "I always speedily got satiated."[94]

So strong was this default repulsion that it determined British attitudes toward ingestion of related species, and in circumstances where more depended on willingness to partake than the enhancement of dietary choice. As, earlier in the century, William Burchell forthrightly regretted in recounting his adventures in the African bush, "I could not . . . resist altogether the misleading influence of prejudice and habit; and allowed myself, merely because I viewed the meat as horseflesh, to reject food which was really good and wholesome." Although he praised the greater rationality of his Hottentot attendants, who ate zebras and quaggas with relish, he apparently communicated to them both the strength of his aversion and its approximate source. Identification with the central tenets of British culture required the renunciation of horsemeat. Thus two Hottentots who accepted baptism soon discovered, notwithstanding their lifelong habit of indulgence, "that they were unable to eat it; . . . that it always created a nausea."[95]

It was, conversely, one of the defining characteristics of foreigners that they craved strange meat, as gorging on the flesh of domesticated ungulates stereotyped Britons. Just across the Channel the French relished frog legs and snails, and in remoter lands still odder tastes further confirmed the transgressive predilections of alien peoples.[96] As Thomas Henry Huxley drily commented after inspecting the fish markets of Rio de Janeiro, "They must eat queer things."[97] Sometimes these preferences were catalogued with scholarly restraint, as contributions to knowledge about either the eater or the eaten, thus conflating eccentric foreigners with the curious creatures they ate as appropriate objects of scrutiny. For example, when describing a new species of jerboa, Thomas Hardwicke noted in passing that "a tribe of low Hindus . . . esteem them good and nutritious food."[98] In his compendium of natural history, William Bingley routinely included such information as that badger flesh, "which is somewhat similar in taste to that of the wild dog, is much esteemed in Italy, France, and Germany," and that "American Indians frequently eat the *flesh* of the skunk."[99] The guide to the national collection of animal products at Bethnal Green specified that "in some countries the flesh of monkeys is eaten"; that "the flesh of a few bats is eaten"; and that the flesh of many rodents "serves for food."[100]

Similar detachment characterized some British attempts to sample nonstandard cuisine. The naturalists on the *Endeavour* voyages extended their research to include the consumption of their specimens; Joseph Banks subsequently claimed that "I have eaten my way into the Animal Kingdom farther than any other man."[101] Even when undertaken in a spirit of bravado and an atmosphere of convivial or even uproarious bonhomie, such ventures nonetheless tended to transpire within a setting recognizably if relaxedly devoted to research. It was thus the pursuit of knowledge that inspired the Oxford geologist William Buckland to regale his family with hedgehog and crocodile and, later, led his son Frank to eat the most appetizing casualties of the Regent's Park zoo.[102] The culinary adventures of the Glutton or Gourmet Club, in the course of which the undergraduate Charles Darwin sampled hawk, bittern, and an old brown owl, developed from more conventional natural history pursuits.[103] The fortnightly dinners of the Raleigh Travellers' Club, forerunner of the Royal Geographical Society, featured specialties from exotic locations that members had visited, while at the convivial (speeches discouraged) gatherings of the Zoological Club of the Zoological Society of London, such eminences as Huxley and William Flower ate kudu and spotted cavy.[104] These institutional frameworks, however tenuously or even satirically embodied, ensured that forays into forbidden territory remained simply exploratory. Even when lightly draped, the protective mantle of zoological investigation guaranteed that there was no danger of going native.

Less restrained and more revealing responses to foreign eating habits ranged from derision to horror. Among some peoples, "even the rat, which is generally held in abhorrence, is said to be wholesome and savoury food."[105] "A Beef Eater," the pseudonymous author of a book purporting to exhibit "the natives of various countries at feeding time," distilled this viewpoint, claiming, for example, that "animals that in England would be looked on with disgust . . . , are by the Chinese regarded as delicacies." The loathsome treats included grubs, earthworms, sphinx moth caterpillars, drowned rats, and, worst of all, domestic dogs and cats.[106] The consumption of companion animals recurred insistently in accounts of exotic eating habits, as both epitome and extreme. A sporting author thus claimed that the flesh of dogs was "preferred by the negroes to that of any other animal," which demonstrated their "unnatural and depraved appetite."[107] According to Bingley, the regrettable habit was widely dispersed: "Disgusting as it may appear to us, the *flesh* of the dog is a favourite food in many countries." Not only was it sold like any other meat in the "markets of Canton," but it was relished in Greenland, North America, and Africa; in the West Indies, on the other hand, he revealed that "the negroes frequently eat the *flesh* of cats."[108] So conventional was the association of the Far East with dog eating that by the late Victorian period "Chinese Edible Dogs," which resembled inelegant reddish pomeranians, were standard fixtures in the "foreign dogs" class at dog shows.[109]

The categories of natural history were sometimes invoked to explain the violent aversion provoked by such practices. It was frequently claimed, for example, that willingness to eat other meat eaters constituted the dividing line between civilization and barbarism. According to a late-nineteenth-century epicure, "certain very low caste people of India and other lands are said to go in for feeding upon carnivorous animals such as dogs, cats, rats, foxes, leopards, wolves, jackals, and other nasty 'brutes' . . . —a diet which must be characterized as simply disgusting."[110] Holt, who recommended only insects that were "strict vegetarians" for human consumption, noted wryly that "carnivorous animals . . . are held unworthy of the questionable dignity of being edible by civilized man."[111] Victorian housekeepers were confidently instructed that "beasts of prey . . . are never employed as food . . . except among savage tribes, or in cases of necessity." Diverse reasons were offered for this anathema. The flesh of such animals was "lean . . . tough . . . coarse and disagreeable"; herbivores were much more "wholesome."[112] Or, the dietary restrictions of the Old Testament forbade the consumption of pawed carnivores, as, indeed, they did of solid-hoofed horses. These prohibitions, however, seldom interfered with Victorian enjoyment of oysters, ham, hare, and lobsters. And as far as members of the order Carnivora were concerned, if the consumption of otters, under the rubric of flesh or of fish, had gone out of fashion long before the nineteenth century, preserved bear meat was still eaten with pleasure when available.[113] But, as formulations that pointedly compared European and exotic practices made clear, the real transgression was a social one, derived from the compact understood to exist between dogs and cats (and horses too) and human beings. By eating these animals, "savages" violated a category properly defined by trust and cooperation.[114] Thus, a domestic encyclopedia lamented that "the *dog*, the faithful companion and friend of man, . . . in many parts of the world . . . is considered as good food."[115]

Such errors could be noted with complacency as well as aversion, as unsurprising if regrettable marks of the inferiority of the people who committed them. It was even suggested that dogs bred for food suffered moral and intellectual degeneration, assuming the same relation to European canines that their masters held to European humans. For example, Charles Darwin claimed that, in comparison with English dogs "valued . . . for their mental qualities and sense, . . . where the dog is kept solely to serve for food, as in the Polynesian islands and China, it is described as an extremely stupid animal."[116] When other Europeans committed similar transgressions they came in for similar, if moderated condescension; it was assumed that they had somehow mistaken what they were eating—but perhaps by not looking too closely where substitution could be anticipated. Prospective British travelers were warned that "hundreds of dogs and cats are annually consumed by the inhabitants of the French capital."[117] More problematic in terms of the hierarchy thus tacitly created was evidence that people of British heritage occasionally relished domestic pets, and not just in a spirit of scientific inquiry. It was reported, for example, that in "that cosmopolitan city San Francisco," restaurants served "bow-wow soup" for 12 cents and "grimalkin steaks" for a quarter.[118] And equivalent outrages occurred much closer to home. In 1771, for example, a "young country lad" ate "a whole cat smothered with onions" at a Cambridge public house, reported in the press as no more or less remarkable than his previous epicurean exploit, which was to devour an eight-pound leg of mutton at a single sitting.[119] Well into the Victorian period cat fanciers unhappily acknowledged that "in some parts of England cats are not wholly despised as an article of diet."[120] A notorious ring of cat eaters operated in West Bromwich, near Birmingham, so efficient in their depredations that "many persons complain that they cannot keep a favourite a week."[121]

The Last Straw

If consuming nonhuman members of the domestic circle was bad, eating people was worse. Cannibalism figured ubiquitously as the essence of savagery. So inextricably intertwined were these attributes that the logic associating them frequently operated in reverse: if all cannibals were savages, then all savages were cannibals. As a result, many of the exotic peoples encountered by British explorers and colonizers were tarred, at least tentatively, with this brush. Fijians were alleged to butcher and bake the extremities before putting their victim to death; indeed, he might be offered a final meal of "his own cooked flesh." One chief was reported to have eaten 900 people.[122] The anthropophagy of the indigenous New Zealanders was endlessly rehearsed, its abandonment a triumph of the imperial mission. The same earthen ovens in which, at the end of the nineteenth century, they cooked potatoes thus "probably, in cannibal times . . . also served for the *long pig* in which they delighted."[123] The Tahitians, too, had been reclaimed by "the humanizing influence of British exertions" from their original position "amongst the most degraded of the human race," practicing "cannibalism and the most barbarous rites."[124] The book of anthropological queries promulgated by the British Association in 1874 to guide "travellers and residents in uncivilized lands" contained a whole section on "Cannibalism," immediately after that on "Food." The first question asked, relatively innocuously, "Does cannibalism prevail?" but the second exposed underlying assumptions: "If it no longer prevails, are there any traditions as to its once having been known?"[125]

Indulgence in human flesh could even have taxonomic consequences, jeopardizing membership in the human species. *Punch* compared "the Manyeuma people" unfavorably with the vegetarian gorillas who lived near them: "Is cannibalism . . . the outcome of a higher degree . . . in the scale of development—a stage more distinctly human?"[126] The motivations most frequently proposed for this behavior made it seem still worse. Hunger was generally dismissed as a possibility, or displaced into the remote past.[127] In 1873, for example, *Nature* reported that the "otherwise gentle Monbuttas" were "habitual cannibals," although they lived among plentiful herds of "elephants, buffaloes, antelopes, and wild swine."[128] Savages thus ate their neighbors and their enemies not to assuage basic needs, which would have been understandable, if not necessarily excusable, but to indulge gratuitous viciousness. In 1831, the London Phrenological Society heard that "in New Zealanders, and other Savage nations . . . the custom did not proceed from hunger, but for the gratification of the revengeful passions."[129] So clearly axiomatic was the low rank of cannibals on the scale of human civilization that even the fact that some cultures reprobated in this way possessed more elevated attributes, including belief in a single deity, was sometimes recorded as cause for puzzlement, but not as reason to reconsider either the implications or the existence of anthropophagy.[130]

Evidence, indeed, was not a conspicuous element of eighteenth- and nineteenth-century discussions of cannibalism. Many adventurers returned to Britain with tales of barbarians who ate conspecifics. As had been the case for centuries if not millennia, however, these accounts derived not from horrified eyewitness but from hearsay and gossip, possibly slander and misinterpretation. And this historical accumulation of anecdote itself increased the likelihood of further corroboration. European travelers who expected uncivilized regions to be stocked with cannibals therefore often found that this was the case.[131] They retailed their accounts to a receptive, even eager public, which similarly conflated barbarism and man-eating. If these attributes did not occur in tandem, it might be necessary to revise other ideas about peoples derogated as "savage," including their relationship to those who were self-consciously "civilized." This dichotomy became increasingly crucial as political relations between these groups intensified. When imperialism entered its most acquisitive and energetic phase after the middle of the nineteenth century, the hunger for stories that would confirm traditional assumptions sharpened. For example, in the middle of *Man's Place in Nature*, an analysis of the relationship between human beings and the great apes, Huxley gratuitously inserted a long extract from a sixteenth-century account of African cannibalism, with an illustration of a butcher's stall stocked with human appendages.[132] Novelists responded in a similar vein, so that the theme of cannibalism assumed an unprecedented prominence in late Victorian fiction.[133]

As in all taxonomies, however, the meaning of the categories was not essential or intrinsic. If cannibalism confirmed the barbarity of exotic peoples, whether they ate human flesh or not, it had no such implications when committed by people incontestably civilized. On the basis of hard evidence, the only people who indisputably qualified as habitual cannibals were of European stock. Travelers who sailed the high seas were liable to shipwreck, and until well into the nineteenth century it was assumed that *in extremis* surviving mariners would kill and eat each other.[134] Explorers marooned in terrestrial wastes routinely made similarly pragmatic calculations.[135] When the remains of the Franklin expedition that had disappeared in quest of the Northwest Passage in 1845 were discovered a decade later, "the mutilated state of many of the bodies & the contents of the kettles . . . [showed] that our wretched countrymen had been driven to the dread alternative of cannibalism."[136] The extent to which such behavior had been regarded as normal and justifiable not only by sailors and adventurers, but by Britons in general, became obvious in 1884, when Captain Tom Dudley and Mate Edwin Stephens of the yacht *Mignonette*, who had been rescued after suffering shipwreck and extreme subsequent exposure, received the death sentence for the murder of one of their shipmates. Their arrest and prosecution greatly surprised the accused malefactors, who had voluntarily told the whole story, killing included, as soon as they had arrived home. The conviction also outraged much of the public, more inclined to treat the brave survivors as heroes than as criminals, so that when Queen Victoria ultimately pardoned them, the *Times* complained of the "mawkish sympathy" they received.[137] The case of *Regina v. Dudley and Stephens* thus marked the beginning of a new legal era, giving notice that, no matter how extenuating their circumstances, future cannibals could no longer assume that British polity was on their side. But, although the decision provoked a great deal of commentary, it was nowhere hailed as marking the emergence of the nation from an existing condition of barbarism.

Even less exigent cannibalism, if it occurred on home ground, was condemned in terms that seemed relatively mild. When cannibalism was announced at St. Thomas's Hospital in London, the *Lancet* reported with relief that the youth who had "cooked and eaten a small piece of human flesh" taken from a corpse was an assistant in the chemical laboratory, rather than one of the medical students, "who are gentlemen by birth and education." But even though he did not belong to this elite group, the erring assistant was not therefore classified among the ranks of savages. He lost his job, but his behavior was otherwise castigated merely as "foolish and indecent."[138] And his eccentric dining habits brought no more generalized disgrace upon his fellow citizens than did those of the cat eaters of West Bromwich.

In a nation self-defined as quintessentially civilized, such lapses inevitably appeared as regrettable eccentricities rather than as representative sins. No matter how many people they ate, the British were never classified as *Homo europaeus anthropophagus*. As one anti-vegetarian had concretely and complacently put it, "those who inhabit the country of roast beef are . . . little in danger of seeing the limbs of their friends exposed to sale in their markets." [139] And if an evolutionary perspective might suggest that in previous millennia "the forefathers of the present European family . . . were not wholly free from one of the worst charges that is laid to savage existence; viz. the practice of cannibalism," such ancient transgressions similarly lacked power to label their refined Victorian descendants.[140]

So axiomatic was the opposition between Britishness and barbarism that it could serve as the basis of humor that also reified the increasingly rigid human taxonomies of the late nineteenth century. When the inner circle of the rigorously racist Anthropological Society constituted itself "The Cannibal Club," using the head of an African to gavel their meetings to order, they had no fear of being confused with their namesakes.[141] The manifest distance between them made the irony obvious.

A cannibal butcher shop.

Notes

1. The Point of Order

1. Joseph Banks, *The Endeavour Journal of Joseph Banks,* ed. J. C. Beaglehole (Sydney: Angus and Robertson, 1962), vol. II, 84; James Cook, *The Journals of Captain James Cook on His Voyages of Discovery: The Voyage of the Endeavour, 1768–1771,* ed. J. C. Beaglehole (Cambridge: Hakluyt Society, Cambridge University Press, 1955), 352.

2. Banks, *Endeavour Journal,* II, 89.

3. Banks, *Endeavour Journal,* II, 93–94; Raphael Cilento, "Sir Joseph Banks and the Kangaroo," *Notes and Records of the Royal Society of London* 26 (1971), 157, 160; Cook, *Journals,* 360.

4. Advertisement in the *Edinburgh Evening Courant,* quoted in Michael A. Taylor, "An Entertainment for the Enlightened; Alexander Weir's Edinburgh Museum of Natural Curiosities, 1782–1802," *Archives of Natural History* 19 (1992), 157.

5. P. J. P. Whitehead, "Zoological Specimens from Captain Cook's Voyages," *Journal of the Society for the Bibliography of Natural History* 5 (1989), 182–184.

6. Harold Burnell Carter, *Sir Joseph Banks, 1743–1820* (London: British Museum (Natural History), 1988), 309–310.

7. William Bingley, *Animal Biography; or, Authentic Anecdotes of the Lives, Manners, and Economy of the Animal Creation* (London: Richard Phillips, 1804) I, 391.

8. See, for example, Zoological Society of London, Minutes of Council, XI, 348–349 (April 1, 1857), in the library of the society; S. J. Woolfall, "History of the 13th Earl of Derby's Menagerie and Aviary at Knowsley Hall, Liverpool (1806–1851)," *Archives of Natural History* 17 (1990), 6, 12.

9. William Henry Flower, "A Century's Progress in Zoological Knowledge," in *Essays on Museums and Other Subjects Concerned with Natural History* (1898; rpt., New York: Books for Libraries Press, 1972), 163.

10. David Collins, *An Account of the English Colony in New South Wales . . .* (1802), quoted in Jacob W. Gruber, "Does the Platypus Lay Eggs? The History of an Event in Science," *Archives of Natural History* 18 (1991), 110–111. On the early reception of platypus specimens, see also Henry Burrell, *The Platypus: Its Discovery,*

Zoological Position, Form and Characteristics (Sydney: Angus and Robertson, 1927), chap. 2.

11. Thomas Bewick, *A General History of Quadrupeds* (Newcastle upon Tyne: T. Bewick, 1824), 523; George Shaw, *General Zoology; or Systematic Natural History* (London: G. Kearsley, 1800), I, 229.

12. Shaw, *General Zoology,* I, 228; Peter Whitehead, *The British Museum (Natural History)* (London: Philip Wilson, 1981), 45.

13. Thomas Brown, *Biographical Sketches and Authentic Anecdotes of Quadrupeds* (Glasgow: A. Fullarton, 1831), 437.

14. "Department of Zoology and Comparative Anatomy Presentation Book. Volume I. 1883–1948," not paginated, Oxford University Museum Archives.

15. George Bennett, "Notes on the Natural History and Habits of the *Ornithorhynchus paradoxus,* Blum.," *Transactions of the Zoological Society of London* 1 (1835), 229, 235.

16. William Swainson, *On the Natural History and Classification of Quadrupeds* (London: Longman, Rees, Orme, Brown, Green and Longman, 1835), 219.

17. Charles Darwin to Philip Parker King, January 21, 1836, in *Correspondence of Charles Darwin,* ed. F. Burkhardt and Sydney Smith (Cambridge: Cambridge University Press, 1985), I, 481.

18. *Punch* 88 (1885), 251.

19. "Donations to the Museum," *Transactions of the Natural History and Antiquarian Society of Penzance* 1 (1851), 19; Acclimatisation Society of Great Britain, and Ornithological Society of London, *Sixth Annual Report* (1866), 64.

20. See Wilma George, "Alive or Dead: Zoological Collections in the Seventeenth Century," 179–187, in *The Origin of Museums: The Cabinet of Curiosities in Sixteenth- and Seventeenth-Century Europe,* ed. Oliver Impey and Arthur MacGregor (Oxford: Clarendon Press, 1985), for a survey of European contact with new animals during this period; Wilma George, "The Bestiary: A Handbook of the Local Fauna," *Archives of Natural History* 10 (1981), 193–195.

21. Harriet Ritvo, *The Animal Estate: The English and Other Creatures in the Victorian Age* (Cambridge, Mass.: Harvard University Press, 1987), 216–217.

22. Robert Kerr, *The Animal Kingdom or Zoological System, of the Celebrated Sir Charles Linnaeus . . .* (London: J. Murray, 1792), 305.

23. Everard Home, "Description of the Anatomy of the *Ornithorhynchus Hystrix,*" *Philosophical Transactions of the Royal Society* 92 (1802), 360. On the early scientific reaction to the echidna, see Jacob W. Gruber, "What Is It? The Echidna Comes to England," *Archives of Natural History* 11 (1982), 1–15, and Stephen Jay Gould, "Bligh's Bounty," *Natural History* 94 (September 1985), 2–10.

24. Busick Harwood, *A Synopsis of a Course of Lectures on the Philosophy of Natural History and the Comparative Structure of Plants and Animals* (Cambridge: Francis Hodson, 1812), 16.

25. John Thomson, "A Description of the Anatomy of the *Ornithorynchus Paradoxus.*

By Everard Home," *Edinburgh Review* 2 (1803), 436. He meant to class the platypus with the reptiles, which he viewed as a subcategory of Amphibia.

26. William MacLeay, "Draft Classification of Mammals on Quinary Principles" (1829), not paginated (manuscript in the Linnean Society Library, London, Case 5A, Ms. 241); William Coulson, "Introduction" to J. F. Blumenbach, *A Manual of Comparative Anatomy*, trans. William Lawrence, rev. William Coulson (London: W. Simpkin and R. Marshall, 1827), xx–xxi.

27. Gruber, "Does the Platypus Lay Eggs?" 57–58. In the twentieth century, in accordance with the rule of priority in zoological nomenclature, *anatinus* has been restored.

28. William Elford Leach, *The Zoological Miscellany; being descriptions of new, or interesting animals* (London: E. Nodder, 1815), II, systematic index.

29. William Bullock, *A Companion to Mr. Bullock's Museum* (London: W. Bullock, 1810), 31–32; John Fleming, *The Philosophy of Zoology* (Edinburgh: Archibald Constable, 1822), II, 211; G. R. Waterhouse, *The Natural History of the Marsupialia or Pouched Animals* (Edinburgh: W. H. Lizars, 1841), 59–60; Thomas Pennant, *History of Quadrupeds* (London: B. and J. White, 1793), II, 31.

30. Edward Turner Bennett, *The Tower Menagerie: Comprising the Natural History of the Animals Contained in that Establishment* (London: Robert Jennings, 1829), 155–156; Cuthbert Collingwood, "On Recurrent Animal Form, and Its Significance in Systematic Zoology," *Annals and Magazine of Natural History,* 3d ser., 6 (1860), 85.

31. H. E. Strickland, "Observations on Classification in Reference to the Essays of Messrs. Jenyns, Newman and Blyth," in William Jardine, *Memoirs of Hugh Strickland, M.A.* (London: John Van Voorst, 1858), 400; Richard Q. Couch, "An Introductory Address on the Study of Natural History," *Transactions of the Natural History and Antiquarian Society of Penzance* 2 (1851), 19.

32. Robert Mudie, *First Lines of Zoology by Question and Answer: For the Use of the Young* (London: Whittaker, Treacher, 1831), 45.

33. Richard Lydekker, *A Hand-book to the Marsupialia and Monotremata* (London: Edward Lloyd, 1896), v.

34. For a modern explanation, see John F. Eisenberg, *The Mammalian Radiations: An Analysis of Trends in Evolution, Adaptation, and Behavior* (Chicago: University of Chicago Press, 1981), chap. 3.

35. Lydekker, *Hand-book to the Marsupialia and Monotremata,* 224–225; St. George Mivart, "On the Possible Dual Origin of the Mammalia," *Proceedings of the Royal Society of London* 93 (1887–88), 372–379; "Mm. H. and E. Milne-Edwards's New Work on Mammals," *Nature* 11 (1875), 463.

36. See, for example, "The New Australian Mammal," *Nature* 42 (1890), 645, and "Dr. Stirling's *Notoryctes typhlops,*" *Nature* 45 (1891), 135.

37. Wilma George, "Sources and Background to Discoveries of New Animals in the Sixteenth and Seventeenth Centuries," *History of Science* 18 (1980), 79–80; Janet

Browne, *The Secular Ark: Studies in the History of Biogeography* (New Haven: Yale University Press, 1983), 16; *"The Zoological Record," Nature* 27 (1883), 310.

38. Whitehead, "Zoological Specimens from Captain Cook's Voyages," 181–183; Adrian Desmond and James Moore, *Darwin* (London: Michael Joseph, 1991), 191, 198.

39. Charles Hamilton Smith, *Introduction to the Mammalia* (Edinburgh: W. H. Lizars, 1846), 76.

40. George Bennett, *Gatherings of a Naturalist in Australasia: Being Observations Principally on the Animal and Vegetable Productions of New South Wales, New Zealand, and Some of the Australian Islands* (London: Jan Van Voorst, 1860), vi.

41. Aristotle, *Historia Animalium*, trans. A. L. Peck (Cambridge, Mass.: Harvard University Press, 1979), I, 33–35; G. E. R. Lloyd, *Science, Folklore and Ideology: Studies in the Life Sciences in Ancient Greece* (Cambridge: Cambridge University Press, 1983), 16; E. S. Russell, *Form and Function: A Contribution to the History of Animal Morphology* (1916; rpt., Chicago: University of Chicago Press, 1982), 3–5.

42. C. E. Raven, *John Ray: Naturalist* (1950; rpt., Cambridge: Cambridge University Press, 1986), 379–380.

43. For nineteenth-century formulations of this position, see Thomas Rhymer Jones, *A General Outline of the Animal Kingdom, and Manual of Comparative Anatomy* (London: John Van Voorst, 1841), 1–2; and "Classification of Mammalia," *Zoological Magazine, or Journal of Natural History* 1 (1833), 177–179.

44. Caleb Hiller Parry to J. Banks, April 26, 1800, in *The Sheep and Wool Correspondence of Joseph Banks, 1781–1820*, ed. Harold B. Carter (New South Wales: The Library Council of New South Wales, 1979), 321.

45. Jones, *General Outline*, 640.

46. Charles Darwin to G. R. Waterhouse, December 3 or 17, 1843, *Correspondence*, II, 415.

47. *A Guide to the Ipswich Museum* (Ipswich: Ipswich Museum, 1871), 20–21.

48. Bewick, *General History of Quadrupeds*, 506, 513.

49. Because the platypus and the echidna lack nipples, scientists did not discover their mechanism for lactation (and thus were not convinced that they lactated) until the 1830s. Gruber, "Does the Platypus Lay Eggs?" 77–80. For a feminist interpretation of Linnaeus's preference for "mammalia," see Londa Schiebinger, *Nature's Body: Gender in the Making of Modern Science* (Boston: Beacon Press, 1993), 42 ff.

50. George, "Sources and Background," 81, 94–95; William B. Ashworth, Jr., "Remarkable Humans and Singular Beasts" in *The Age of the Marvelous*, ed. Joy Kenseth (Hanover, N.H.: Hood Museum of Art, Dartmouth College, 1991), 124–125.

51. M. F. Ashley Montagu, *Edward Tyson, M.D., F.R.S., and the Rise of Human and Comparative Anatomy in England* (Philadelphia: American Philosophical Society, 1943), 353–354.

52. Charles Darwin, *Correspondence*, II, 377–383, 415–416.

53. Adrian Desmond, *The Politics of Evolution: Morphology, Medicine, and Reform in Radical London* (Chicago: University of Chicago Press, 1989), 278–288; Gruber, "Does the Platypus Lay Eggs?" 74–86.

54. "The Royal Society and Anatomy of the Platypus," *Lancet* 2 (1840), 105.
55. Everard Home to W. S. MacLeay, n.d. (1825), MacLeay Correspondence, Linnean Society Library, London.
56. Burrell, *Platypus,* chap. 4, esp. p. 45; Desmond, *Politics of Evolution,* 178.
57. "The South Australian Court at the Colonial Exhibition," *Illustrated London News* 88 (1886), 578.
58. This interpretation has proved extremely durable. Modern interpreters with widely differing scholarly preoccupations and equally diverse assessments of the social and intellectual consequences of this development have echoed the traditional zoological consensus. According to art historian Barbara Stafford, the eighteenth century was "the great age of classification"; *Voyage into Substance: Art, Nature, and the Illustrated Travel Account, 1760–1840* (Cambridge, Mass.: MIT Press, 1987), 54. According to Michel Foucault, it was a time when "Natural History was . . . above all . . . a set of rules for arranging statements in series, an obligatory set of schemata of dependence, or order, and of succession"; *The Archaeology of Knowledge,* trans. A. M. Sheridan Smith (New York: Harper and Row, 1976), 57. In the view of the cultural historian Keith Thomas, the systems so constituted embodied "a novel way of looking at things, . . . which was more detached, more objective, less man-centered than that of the past"—they were distinctive, path-breaking, and significant; *Man and the Natural World* (New York: Pantheon, 1983), 52. Nor did the fact that they were ultimately superseded or modified—by taxonomies that could be characterized, depending on the inclinations of the analyst, as more historical (Michel Foucault, *The Order of Things* [New York: Vintage, 1973], xxii–xxiii), more anatomical (M. J. Novacek, "Characters and Cladograms: Examples from Zoological Systematics," in *Biological Metaphor and Cladistic Classification,* ed. H. M. Hoenigswald and L. F. Wiener [Philadelphia: University of Pennsylvania Press, 1987], 190–191), or more natural (Mary P. Winsor, *Starfish, Jellyfish, and the Order of Life: Issues in Nineteenth-Century Science* [New Haven: Yale University Press, 1976], 1)—compromise their privileged historical position. Instead, in some whiggish versions of the history of biology, the introduction of systematic taxonomy has been viewed as the opening battle in a protracted triumph of human intelligence over nature and, within the human realm, of science over superstition.
59. Carl von Linné, *A General System of Nature . . . ,* ed. and trans. William Turton (London: Lackington, Allen, 1806), I, vi.
60. Kerr, *Animal Kingdom,* v.
61. Linnaeus, *Miscellaneous Tracts Relating to Natural History,* trans. and ed. Benjamin Stillingfleet (London: R. J. Dodsley, 1759), xx–xxi.
62. William Borlase, *Natural History of Cornwall* (Oxford: W. Jackson, 1768), viii; Richard Pulteney, *A General View of the Writings of Linnaeus* (London: J. Mawman, 1805), 11.
63. William Jardine, "Memoir of Aristotle," in *The Natural History of Gallinaceous Birds, Part I* (Edinburgh: W. H. Lizars, 1834), 19.
64. These sketches preface the following volumes of the Naturalist's Library (respec-

tively): James Duncan, *The Natural History of Beetles* (Edinburgh: W. H. Lizars, 1835); William Jardine, *Humming-Birds*, vol. I (Edinburgh: W. H. Lizars, 1833); William Jardine, *Humming-Birds*, vol. II (Edinburgh: W. H. Lizars, 1833); and William Jardine, *The Natural History of the Nectariniadae, or Sun-Birds* (Edinburgh: W. H. Lizars, 1843).

65. Jardine, *Humming-Birds*, I, 26.

66. [W. Kirkland (?)], "The Museum" (printed but no publication data), pasted in "Book of Benefactions—1757 to 1769. 1824 to 1829," Ashmolean Museum Archives, Oxford.

67. Edward Tyson, *Phocaena, or the Anatomy of a Porpesse* (1690), quoted in Ashley Montagu, *Edward Tyson*, 95.

68. Pennant, *History of Quadrupeds*, I, i.

69. Richard Brookes, *New and Accurate System of Natural History* (London: J. Newbery, 1763), I, x–xi.

70. Charles Hamilton Smith, "Memoir of Gesner," in *The Natural History of Horses* (Edinburgh: W. H. Lizars, 1841), 42; Andrew Crichton, "Memoir of Pliny," in Prideaux John Selby, *The Natural History of Gallinaceous Birds, Part III. Pigeons* (Edinburgh: W. H. Lizars, 1834), 74; Smith, *Introduction to Mammalia*, 75.

71. Richard Owen, "Mammalia," in *The Cyclopedia of Anatomy and Physiology*, ed. Robert Bentley Todd (London: Sherwood, Gilbert, and Piper, 1836–1859), III, 237.

72. H. B. Tristram, "Recent Geographical and Historical Progress in Zoology," *Contemporary Review* 2 (1866), 104; William Houghton, "Aristotle's *History of Animals*," *Quarterly Review* 117 (1865), 57. Philosophical opinions about the extent of Aristotle's influence on Enlightenment (and subsequent) taxonomy continue to differ. For example, David L. Hull has argued for its continuing strength in *Science as Process: An Evolutionary Account of the Social and Conceptual Development of Science* (Chicago: University of Chicago Press, 1988), chap. 3. For a summary of recent views see Juliet Clutton-Brock, "Aristotle, the Scale of Nature, and Modern Attitudes to Animals," *Social Research* 62 (1995), 421–440.

73. Leonard Chappelow, "The Sentimental Naturalist, Containing a Series of Descriptive Pictures, Taken from the Vegetable and Animal Kingdoms, The Latter Arranged According to the Celebrated System of Linnaeus" (ca. 1809), I, 17, ll. 408–414. Manuscript in Cambridge University Library, Na.1.33.

74. Thomas Pennant, *Arctic Zoology* (London: Henry Hughes, 1784), 3. The British were far from alone in their conflation of taxonomy and empire. For a view of related developments, focused on Spanish America, see Mary Louise Pratt, *Imperial Eyes: Travel Writing and Transculturation* (New York: Routledge, 1992), 24–37.

75. For accounts of Linnaeus's reception outside of Britain, see, for example, James L. Larsen, *Interpreting Nature: The Science of Living Form from Linnaeus to Kant* (Baltimore: Johns Hopkins University Press, 1994); Phillip R. Sloan, "The Buffon-Linnaeus Controversy," *Isis* 67 (1976), 356–375; and Henri Daudin, *De Linné à Lamarck: Méthodes de la Classification et Idée de Serie en Botanique et en Zoologie (1740–1790)* (Paris: Librairie Felix Alcan, 1926).

76. For example, Charles Hamilton Smith considered "the Pachydermata . . . less uniform in general character" than other mammalian orders (*Introduction to Mammalia*, 268), and William Swainson characterized the hyenas as "an isolated group" (*On the Natural History and Classification of Quadrupeds*, 131).

77. On the spread of amateur interest in natural history, see David Elliston Allen, *The Naturalist in Britain: A Social History* (Harmondsworth, Middlesex: Penguin, 1978), chaps. 1–4.

78. On zoological works for juvenile audiences, see Harriet Ritvo, "Learning from Animals: Natural History for Children in the Eighteenth and Nineteenth Centuries," *Children's Literature* 14 (1985), 72–93.

79. Edward Topsell, *The Historie of Foure-Footed Beastes* (London: William Iaggard, 1607), 102–107.

80. Topsell, *Historie of Foure-Footed Beastes*, 711–721.

81. Ritvo, *Animal Estate*, 12–13. William B. Ashworth has argued persuasively that contemporaries perceived coherence rather than chaos in "Emblematic Natural History of the Renaissance," in *Cultures of Natural History*, ed. N. Jardine, J. A. Secord, and E. C. Spary (Cambridge: Cambridge University Press, 1996), 17–37.

82. Bewick, *General History of Quadrupeds*, iii, v–x.

83. Bewick, *General History of Quadrupeds*, 231–234.

84. Pennant, *History of Quadrupeds*, I, 295–297.

85. Arthur MacGregor, "The Cabinet of Curiosities in Seventeenth-Century Britain," in Impey and MacGregor, eds., *Origin of Museums*, 151–152; Raven, *John Ray*, 329.

86. John Tradescant, *Museum Tradescantianum: or, A Collection of Rarities Preserved At South-Lambeth neer London* (1656; rpt., Oxford: Old Ashmolean Reprints, 1925), not paginated; see R. F. Ovenell, *The Ashmolean Museum, 1683–1894* (Oxford: Clarendon Press, 1986), for an account of the transfer of the collection.

87. John Church, *A Cabinet of Quadrupeds* (London: Darton and Harvey, 1805), I, not paginated.

88. G. L. Leclerc, Comte de Buffon, *Natural History . . . with Occasional Notes . . . by the Translator* (London: W. Strahan and T. Cadell, 1781–85), III, 5; VIII, 287–301.

89. [James] Duncan, *Introduction to the Catalogue of the Ashmolean Museum* (n.p., ca. 1830), 25–26, 31–41, 50.

90. Carter, *Sir Joseph Banks*, 46; James Anderson, *Recreations in Agriculture, Natural History, Arts, and Miscellaneous Literature* (London: T. Bentley, 1799), I, 2. For a related recent perspective, see Mario A. DiGregorio, "In Search of the Natural System: Problems of Zoological Classification in Victorian Britain," *History and Philosophy of the Life Sciences* 4 (1982), 225.

91. John Berkenhout, *Synopsis of the Natural History of Great Britain and Ireland* (London: T. Cadell, 1795), I, v.

92. Borlase, *Natural History of Cornwall*, ix.

93. Kerr, *Animal Kingdom*, 1.

94. Jardine, *Humming-Birds*, I, 27–28.

95. Georges Louis Leclerc, Comte de Buffon, *Barr's Buffon: Buffon's Natural History*

(London: H. D. Symonds, 1797), VII, 53. Thomas Pennant expressed similar, although more moderate, sentiments in *History of Quadrupeds,* I, iii.

96. "An Essay on a Method of Classing Animals," *Annual Register* (1759), quoted in A. J. Cain, "Natural Classification," *Proceedings of the Linnean Society of London* 174 (1963), 117, 119.

97. Swainson, *On the Natural History and Classification of Quadrupeds,* 34; John Fleming, *The Philosophy of Zoology; or A General View of the Structure, Functions and Classification of Animals* (Edinburgh: Archibald Constable, 1822), I, vi–vii.

98. Dr. David Skene to John Ellis, October 6, 1768, in Spencer Savage, ed., *Catalogue of the Manuscripts in the Library of the Linnean Society of London. Part IV.— Calendar of the Ellis Manuscripts* (London: Linnean Society, 1948), 30. See also John Lyon, "The 'Initial Discourse' to Buffon's *Histoire naturelle:* The First Complete English Translation," *Journal of the History of Biology* 9 (1976), 133–181.

99. Buffon, *Barr's Buffon,* VII, 198.

100. Oliver Goldsmith, *An History of the Earth, and Animated Nature* (London: J. Nourse, 1774), III, 2–3; IV, 100.

101. *The Menageries: The Natural History of Monkeys, Opossums, and Lemurs* (London: Charles Knight, 1838), 9.

102. See Richard Yeo, "Reading Encyclopedias: Science and the Organization of Knowledge in British Dictionaries of Arts and Sciences, 1730–1850," *Isis* 82 (1991), 28–29, and Wilda Anderson, *Between the Library and the Laboratory: The Language of Chemistry in Eighteenth-Century France* (Baltimore: Johns Hopkins University Press, 1984), 41–42. Bestiaries were often organized (both in Latin and in translation) alphabetically, according to the Latin names of the animals.

103. William Frederic Martyn, *A New Dictionary of Natural History; or, Compleat Universal Display of Animated Nature* (London: Harrison, 1785), I, not paginated.

104. *A Dictionary of Natural History; or Complete Summary of Zoology* (London: C. Whittingham, 1802), iii.

105. John Bigland, *Letters on Natural History, Exhibiting a View of the Power, Wisdom, and Goodness of the Deity* (London: Longman, Hurst, Rees and Orme, 1806), iv–v.

106. *Popular Zoology: Comprising Memoirs and Anecdotes of the Quadrupeds, Birds, and Reptiles, in the Zoological Society's Menagerie* (London: John Sharpe, 1832), viii.

107. Martyn, *New Dictionary,* vols. I and II (not paginated).

108. Cymmrodorion Society, *The British Zoology* (London: J. and J. March, 1766), 48; Duncan, *Natural History of Beetles,* 17; [John Fleming], "On Systems and Methods in Natural History. By J. E. Bicheno," *Quarterly Review* 41 (1829), 303–304.

109. John Walker Lectures, "Dr. Walker's Lectures, July 1791" (notes taken by students), Edinburgh University Library Special Collections Ms. Dc.7.113, vol. III.

110. Brookes, *New and Accurate System,* I, xi.

111. Gilbert White, *The Natural History of Selborne,* ed. Richard Mabey (Harmondsworth, Middlesex: Penguin, 1977), 136.

112. William Wood, *Zoography; or the Beauties of Nature Displayed* (London: Cadell and Davies, 1807), I, xvii.

113. Berkenhout, *Synopsis of the Natural History of Great Britain and Ireland*, I, xi.

114. William Holloway and John Branch, *The British Museum; or, Elegant Repository of Natural History* (London: John Badcock, 1803), I, iv.

115. Ernst Mayr has pointed out the paradoxical increase of parochialism within national or linguistic scientific communities after the Renaissance—in his view due to the disappearance of Latin as the universal language of scholarship—at the same time that the machinery of modern international intellectual exchange was being developed; *The Growth of Biological Thought: Diversity, Evolution, and Inheritance* (Cambridge, Mass.: Harvard University Press, 1982), 109–110.

116. William Lawrence, *Lectures on Comparative Anatomy, Physiology, Zoology and the Natural History of Man* (London: R. Carlile, 1823), 68; Smith, *Introduction to Mammalia*, 76.

117. For accounts of the debate over artificial and natural systems of classification in the eighteenth and nineteenth centuries, see David Knight, *Ordering the World: A History of Classifying Man* (London: Burnett Books, 1981), 57–119; DiGregorio, "In Search of the Natural System," 225–240; Winsor, *Starfish, Jellyfish, and the Order of Life*, 1–3; and Paul Farber, *The Emergence of Ornithology as a Scientific Discipline: 1760–1850* (Dordrecht, Holland: D. Reidel, 1982), 79–81.

118. David Elliston Allen, "The Natural History Society in Britain through the Years," *Archives of Natural History* 14 (1987), 245–246; David Elliston Allen, *The Botanists: A History of the Botanical Society of the British Isles through a Hundred and Fifty Years* (London: St. Paul's Bibliographies, 1986), 4–5.

119. J. E. Bicheno, "On Systems and Methods in Natural History," *Linnean Society of London, Transactions* 15 (1827), 479.

120. William S. MacLeay, *Horae Entomologicae: or Essays on the Annulose Animals* (London: S. Bagster, 1819), I, viii.

121. James Rennie, *Alphabet of Zoology for the Use of Beginners* (London: Orr and Smith, 1833), 135. The term in the title was not meant literally; Rennie organized his material analytically.

122. Darwin, *Correspondence*, II, 375–376.

123. Fleming, "On Systems and Methods," 314–315.

124. Elizabeth Gaskell, *Mary Barton: A Tale of Manchester Life* (London: J. M. Dent, 1911), 34.

125. Anne Secord, "Science in the Pub: Artisan Botanists in Early Nineteenth-Century Lancashire," *History of Science* 32 (1994), 269–315.

126. The chain of being derived ultimately from the work of Aristotle (Hull, *Science as Process*, 82). On the chain generally, see Arthur O. Lovejoy, *The Great Chain of Being: A Study of the History of an Idea* (Cambridge, Mass.: Harvard University Press, 1936), esp. chap. 8.

127. Delabère Blaine, *The Outlines of Veterinary Art* (London: A. Strahan, 1802), I, 141, 144–145.

128. Bewick, *General History of Quadrupeds*, 74; Bennett, *Tower Menagerie*, 61; John Lawrence, *A General Treatise on Cattle* (London: Sherwood, Gilbert, and Piper, 1808), 425–426; Pennant, *History of Quadrupeds*, II, 252.

129. Thomas Beddoes, "On the Chain of Being," Minutes, Society for Investigating Natural History, IV, 139–140. Manuscript in Edinburgh University Library Special Collections.

130. Borlase, *Natural History of Cornwall*, viii.

131. Charles White, *An Account of the Regular Gradation in Man, and in Different Animals and Vegetables; and from the Former to the Latter* (London: C. Dilly, 1799), 39.

132. W. D. Ian Rolfe, "William and John Hunter: Breaking the Great Chain of Being," in W. F. Bynum and Roy Porter, eds., *William Hunter and the Eighteenth-Century Medical World* (Cambridge: Cambridge University Press, 1985), 316–318.

133. "Solitarius," "Remarks upon Zoological Nomenclature and Systems of Classification," *Field Naturalist* 1 (1833), 526; Charles Darwin, *Charles Darwin's Notebooks, 1836–1844: Geology, Transmutation of Species, Metaphysical Inquiries*, transcribed and ed. Paul H. Barrett, Peter J. Gautrey, Sandra Herbert, David Kohn, and Sydney Smith (Ithaca, N.Y.: Cornell University Press, 1981), 308.

134. Hull, *Science as Process*, 85; Goldsmith, *History of the Earth*, IV, 187.

135. Hugh Strickland, "On the True Method of Discovering the Natural System in Zoology and Botany," *Annals and Magazine of Natural History* 6 (1841), 192.

136. [Michael Foster], "Higher and Lower Animals," *Quarterly Review* 127 (1869), 400.

137. Alexander MacAlister, *An Introduction to the Systematic Zoology and Morphology of Vertebrate Animals* (Dublin: Hodges, Foster and Figgis, 1878), iv–v.

138. Montagu Browne, *Practical Taxidermy: A Manual of Instruction to the Amateur in Collecting, Preserving and Setting Up Natural History Specimens of All Kinds* (London: L. Upcott Gill, 1884), 320.

139. Matthew M. Milburn, *Sheep and Shepherding* (London: William S. Orr, 1853), 40; W. Lauder Lindsay, "Community of Disease in Man and Other Animals," offprint from *Medical-Chirurgical Review* (1872), 3.

140. Robert Knox, *The Races of Men: A Philosophical Enquiry into the Influence of Race over the Destinies of Nations* (London: Henry Renshaw, 1862), 503. It is possible that this adoption of an outmoded metaphor was semi-intentional, since the rejection of the chain was associated with the late-eighteenth-century defense of monogenism (the idea that the entire human race is of one species) by British scientists, a position that Knox wished to counter. Nancy Stepan, *The Idea of Race in Science: Great Britain, 1800–1960* (Hamden, Conn.: Archon Books, 1982), 6.

141. William Henry Flower, *The Horse: A Study in Natural History* (London: Kegan, Paul, Trench, Trübner, 1891), 9.

142. Recent discussions of the quinary system include Adrian Desmond, "The Making of Institutional Zoology in London, 1822–1836: Part I," *History of Science* 23 (1985), 160–164; Knight, *Ordering the World*, 93–105; Winsor, *Starfish, Jellyfish, and the Order of Life*, 82–87; DiGregorio, "In Search of the Natural System," 234–236; Alec

L. Panchen, *Classification, Evolution, and the Nature of Biology* (Cambridge: Cambridge University Press, 1992), 25–30; and Hull, *Science as Process,* 93–96.

143. On the variety of graphic representations inspired by nineteenth-century debates about the signficance of similarities among diverse organisms, see Robert J. O'Hara, "Telling the Tree: Narrative Representation and the Study of Evolutionary History," *Biology and Philosophy* 7 (1992), 135–160, and "Representations of the Natural System in the Nineteenth Century," *Biology and Philosophy* 6 (1991), 255–274.

144. MacLeay, *Horae Entomologicae,* I, xxv.

145. A few years before MacLeay propounded his system, the German anatomist Lorenz Oken had also elaborated a natural system based on the number five, but apparently the two were not connected. Stephen Jay Gould, "The Rule of Five," *Natural History* 93 (October 1984), 14–23. A more extravagant and eccentric numerological system, based on such quantities as the number of petals and stamens in plants, the number of ribs and vertebrae in animals, and their squares and multiples, was elaborated by Richard Vyvyan in *An Essay on Arithmo-Physiology, or a New Chronological Classification of Organized Matter; Deduced from an Inspection of the Numbers Employed in the Divisions or Subdivisions of the Different Parts of Vegetables and Animals* (London: John Nichols, 1825). Unlike quinarian theory, however, Vyvyan's work had no discernible influence.

146. William S. MacLeay, "Draft Classification of Mammals on Quinary Principles," not paginated.

147. Zoological Club of the Linnean Society, Minute Book 1823–1829. Ms. Drawer 36, Linnean Society Library.

148. Zoological Club Minute Book, 1823–1829, November 30, 1829, not paginated.

149. On Swainson, see Paul Lawrence Farber, "Aspiring Naturalists and Their Frustrations: The Case of William Swainson (1789–1855)," in Alwyne Wheeler and James H. Price, eds., *From Linnaeus to Darwin: Commentaries on the History of Biology and Geology* (London: Society for the History of Natural History, 1985), 51–59; and David Knight, "William Swainson: Naturalist, Author and Illustrator," *Archives of Natural History* 13 (1986), 275–290.

150. Fleming, "On Systems and Methods," 302–303, 322; John Fleming, *A History of British Animals, Exhibiting the Descriptive Characters and Systematical Arrangement of the Genera and Species of Quadrupeds, Birds, Fishes, Mollusca, and Radiata of the United Kingdom* (Edinburgh: Bell and Bradfute, 1828), xxi.

151. Jones, *General Outline,* vii.

152. Darwin, *Correspondence,* III, 109n; II, 416; Darwin, *Charles Darwin's Notebooks,* 354. Adrian Desmond has pointed out that quinarianism could be used by its politically conservative advocates to combat the Lamarckian transformism or evolutionism endorsed by politically radical zoologists, who were based in comparative anatomy rather than descriptive natural history. "Making of Institutional Zoology in London, 1822–1826: Part I," 163.

153. H. N. Turner, "An Essay on Classification," *Zoologist* 5 (1847), 1945.

154. Turner, "Essay on Classification," 1945.

155. "Classification of Mammalia," *Zoological Magazine, or Journal of Natural History* 1 (1833), 176; J. E. Gray, "An Outline of an Attempt at the Disposition of Mammalia into Tribes and Families," *Annals of Philosophy* 26 (1825), 344.

156. Leonard Jenyns, "Report on the Recent Progress and Present State of Zoology," *Report of the Fourth Meeting of the British Association* (London: John Murray, 1835), 155–156.

157. Owen, "Mammalia," in Todd, ed., *Cyclopedia of Anatomy and Physiology*, III, 242. He followed this statement with several pages of direct quotation from MacLeay.

158. Adrian Desmond, *Huxley: The Devil's Disciple* (London: Michael Joseph, 1994), 89–90.

159. J. C. Loudoun, *An Encyclopedia of Agriculture* (London: Longman, Rees, Orme, Brown, Green, and Longman, 1835), 235; Leonard Jenyns [Blomefield], *A Manual of British Vertebrate Animals* (Cambridge: Pitt Press, 1835) I, xvi.

160. Robert Chambers, *Vestiges of the Natural History of Creation and Other Evolutionary Writings*, ed. James A. Secord (Chicago: University of Chicago Press, 1994), 236–273.

161. F. W. Rudler, "On Natural History Museums," in *Papers Relating to the Proposal for the Establishment of a Local Museum in Essex* (Buckhurst Hill, Essex: Essex Field Club, 1891), 15. Rudler's remarks had originally been presented to the Cymmrodorion Society in 1876.

162. John Ruskin, *Love's Meinie: Lectures on Greek and English Birds* (1881), in E. T. Cook and Alexander Wedderburn, eds., *The Works of John Ruskin* (London: George Allen, 1906), XXV, 112.

163. "Zoological Nonsense," *Nature* 12 (1875), 128.

164. Everard Home, *Lectures on Comparative Anatomy; in which are explained the Preparations in the Hunterian Collection* (London: G. and W. Nicol, 1814–1820), I, 6–7; Flower, "The Museum of the Royal College of Surgeons of England" in *Essays on Museums*, 86. For a modern description and analysis of Hunter's system of arrangement, see Stephen J. Cross, "John Hunter, the Animal OEconomy, and Late Eighteenth-Century Physiological Discourse," *Studies in the History of Biology* 5 (1981), 13–21.

165. Cross, "John Hunter," 20.

166. Nehemiah Grew, *Comparative Anatomy of Stomachs and Guts* (1681), quoted in Francis Joseph Cole, *A History of Comparative Anatomy: From Aristotle to the Eighteenth Century* (London: Macmillan, 1944), 246.

167. John Walker Lectures, ca. 1791, Edinburgh University Library Special Collections Ms. Dc.7.113, vol. III, 4; George Graves, *The Naturalist's Companion, Being a Brief Introduction to the Different Branches of Natural History* (London: Longman, Hurst, Rees, Orme, Brown, and Green, 1824), 13.

168. Lavater, Sue and Co. [pseud.], *Lavater's Looking-Glass; or, Essays on the Face of Animated Nature, from Man to Plants* (London: Millar Ritchie, 1800), 167;

Wernerian Natural History Society Minute Books 1808–1858, I, 88 (April 15, 1816). Edinburgh University Library Special Collections Ms. Dc. 2.55–56.

169. C. L. Bonaparte, "A New Systematic Arrangement of Vertebrated Animals," *Linnean Society of London, Transactions* 18 (1837), 305–306; Richard Owen, "On the Characters, Principles of Division, and Primary Groups of the Class MAMMALIA," *Journal of the Proceedings of the Linnean Society of London* 2 (1858), 13.

170. Kerr, *Animal Kingdom,* 36.

171. See, for example, Everard Home, "An Arrangement of the Animal Kingdom Founded on the Modifications of the Egg," in his *Lectures on Comparative Anatomy,* III, 461–74; and P. H. Pye-Smith, "The Placental Classification of Mammals," *Nature* 5 (1872), 381–382.

172. T. H. Huxley, *On Zoology* (London: HMSO, 1869), 4. As Adrian Desmond has illustrated in *The Politics of Evolution,* the antagonism between these two groups reflected a range of opposing commitments, political as well as scientific.

173. John Ruskin, *The Eagle's Nest: Ten Lectures on the Relations of Natural Science to Art* (Orpington, Kent: George Allen, 1880), 156.

174. Lawrence, *Lectures on Comparative Anatomy,* 46–47.

175. Home, *Lectures on Comparative Anatomy,* I, 6–7.

176. Thomas, *Man and the Natural World,* 70–81; Agnes Arber, *Herbals, Their Origin and Evolution: A Chapter in the History of Botany, 1470–1670* (Cambridge: Cambridge University Press, 1938), 166–171.

177. The relationship between folk classification and scientific taxonomy has been frequently discussed by historians and anthropologists. Many have been inclined to emphasize the shared observational or commonsense foundations of both enterprises. See, for example, Leon Croizat, "History and Nomenclature of the Higher Units of Classification," *Bulletin of the Torrey Botanical Club* 72 (1945), 52–53; Lloyd, *Science, Folklore and Ideology,* 3; and Scott Atran, *Cognitive Foundations of Natural History: Towards an Anthropology of Science* (Cambridge: Cambridge University Press, 1990), esp. chap. 9. In a similar vein, Paul Barber has suggested that folklorists exaggerate the difference between folklore and science by preferentially recording folk convictions at odds with their own beliefs. *Vampires, Burial, and Death: Folklore and Reality* (New Haven: Yale University Press, 1988), 27. Brent Berlin has attempted to identify a universal system of classification underlying all folk systems and linking them to scientific taxonomy. "The Relations of Folk Systematics to Biological Classification," *Annual Review of Ecology and Systematics* 4 (1973), 259–271.

On the other hand, scholars have claimed that folk and scientific systems are linked not by consistent observation of the world of nature but by a shared human tendency to perceive that world in terms of social categories. See, especially, Mary Douglas, "The Pangolin Revisited: A New Approach to Animal Symbolism," in Roy Willis, ed., *Signifying Animals: Human Meaning in the Natural World* (London: Unwin Hyman, 1989), 25–36, and *Purity and Danger: An Analysis of the Concepts of Pollution and Taboo* (London: Routledge and Kegan Paul, 1984); see also, for example, Denis Chevallier, *L'Homme, Le Porc, L'Abeille et Le Chien: La Relation Homme-Animal dans le Haut-Diois* (Paris: Institut d'Ethnologie, 1987), 180–194; and Eugene Hunn, "The Utilitarian Factor in Folk Biological Classification," *American Anthropologist* 84 (1982), 830–847.

178. Nicholas Cox, *The Gentleman's Recreation* (London: Thomas Fabian, 1677), I, 121; Pennant, *History of Quadrupeds*, I, 269.

179. Because of this propensity, David Elliston Allen has characterized natural history as having "its flank lying open to the lay world [where] science is . . . at its most porous emotionally," vulnerable to taste and fashion; he has further asserted that "such influences exert a distorting effect on scientific development that is much underrated." "Natural History and Social History," *Journal of the Society for the Bibliography of Natural History* 7 (1976), 510.

180. Brookes, *New and Accurate System*, I, xi.

181. Brookes, *New and Accurate System*, I, xxvi.

182. George Henry Millar, *A New, Complete, and Universal Body, or System of Natural History* (London: Alexander Hogg, 1785), 5.

183. Shaw, *General Zoology*, I, 397.

184. Buffon, *Barr's Buffon*, VII, 24.

185. [J. Gregory], *A Comparative View of the State and Faculties of Man with those of the Animal World* (London: J. Dodsley, 1772), 13.

186. Cecil [Cornelius Tongue], *Hints on Agriculture, Relative to Profitable Draining and Manuring; also the Comparative Merits of the Pure Breeds of Cattle and Sheep* (London: Thomas Cautley Newby, 1858), 118.

187. Swainson, *On the Natural History and Classification of Quadrupeds*, 137.

188. Goldsmith, *History of the Earth*, II, 302.

189. Buffon, *Barr's Buffon*, VI, 13.

190. William Jardine, *The Natural History of the Ruminating Animals, Part I* (Edinburgh: W. H. Lizars, 1835), 133.

191. William Jardine, *The Natural History of the Ruminating Animals, Part II* (Edinburgh: W. H. Lizars, 1836), 236.

192. Pennant, *History of Quadrupeds*, I, 295. Questions about the classification of this familiar domestic animal have only recently been settled, if at all. See Juliet Clutton-Brock, *The British Museum Book of Cats: Ancient and Modern* (London: British Museum Publications, 1988), 6–13, and James Serpell, "The Domestication and History of the Cat," 151, in *The Domestic Cat: The Biology of Its Behavior*, ed. Dennis C. Turner and Patrick Bateson (Cambridge: Cambridge University Press, 1988), for authoritative discussions; see the other contributions to *The Domestic Cat* for examples of inconsistent usage.

193. Bewick, *General History of Quadrupeds*, 1–23; Church, *Cabinet of Quadrupeds*, II, n.p.

194. Shaw, *General Zoology*, I, 277–280.

195. Bennett, *Tower Menagerie*, 85–86. For an account of contemporary French strug-

gles with the scientific classification of domestic animals, see Jean-Pierre Digard, *L'homme et les animaux domestiques: Anthropologie d'une passion* (Paris: Fayard, 1990), chap. 4.

196. On the general overlap of political and zoological categories, see Janet Browne, "A Science of Empire: British Biogeography before Darwin," *Review of the History of Science* 45 (1992), 451–475.

197. Thomas Rowlandson, *Foreign and Domestic Animals Drawn from Nature by Gilpin, Catton, &c. and Etch'd by Thomas Rowlandson* (London: Thomas Rowlandson, 1787).

198. Tyson, *Phocaena,* quoted in Ashley Montagu, *Edward Tyson,* 95.

199. George Johnston, *Prospectus of the Ray Society* (1843); Strickland to Jardine, December 19, 1843, quoted in William Jardine, *Memoirs of Hugh Edwin Strickland, M.A.* (London: John Van Voorst, 1858), ccxii.

200. Richard Lydekker, *A Hand-Book to the British Mammalia* (London: W. H. Allen, 1895), vii; *British Naturalist* 1 (1891), i.

201. Herbert Norris, *History of St. Ives* (St. Ives: Hunts County Guardian, 1889), 82; Oxley Graham, "Mammalia," in *Historical and Scientific Survey of York and District,* ed. George A. Auden (York: John Sampson, 1906), 314.

202. Quoted in Taylor, "An Entertainment for the Enlightened," 156.

203. *The London Museum, and Institute of Natural History,* 1807, John Johnson Collection (Museums 1), Bodleian Library, Oxford.

204. G. R. Waterhouse, *Catalogue of the Mammalia Preserved in the Museum of the Zoological Society of London* (London: Richard and John Taylor, 1838), 2.

205. Albert Günther, *The History of the Collections Contained in the Natural History Departments of the British Museum,* II: *Appendix. General History of the Department of Zoology from 1856 to 1895* (London: British Museum, 1912), 61.

206. William Elford Leach, *Systematic Catalogue of the Indigenous Mammalia and Birds in the British Museum,* ed. Osbert Salvin (1816; London: Willughby Society, 1882), 3, 43.

207. Flower, "School Museums," in *Essays on Museums,* 60–61; J. S. Henslow, *On Typical Series of Objects in Natural History Adapted to Local Museums* (London: Richard Taylor and William Francis, 1856), 5.

208. *Worth's Exeter Cathedral and City Guide Book* (ca. 1880), 98; William Rees, *Cardiff: A History of the City* (Cardiff: The Corporation of the City of Cardiff, 1969), 326–327.

209. *An Abridged Catalogue of the Saffron Walden Museum* (Saffron Walden, 1845), iii.

210. Fleming, *A History of British Animals,* v.

211. Günther, *History of the Collections,* 61.

212. Robert Patterson, *Introduction to Zoology, for the Use of Schools* (London: Simms and M'Intyre, 1857), iii.

213. For a comprehensive account of the development of biogeography, see Browne, *Secular Ark;* for a more restricted analysis, see Gareth Nelson, "From Candolle to Croizat: Comments on the History of Biogeography," *Journal of the History of*

Biology 11 (1978), 269–305. Browne has explored the imperial dimension of this discipline in "Biogeography and Empire," in *Cultures of Natural History,* ed. Jardine, Secord, and Spary, 305–321.

214. See, for example, H. B. Tristram, "Recent Geographical and Historical Progress in Zoology," *Contemporary Review* 2 (1866), 106, and P. L. Sclater, "On the Present State of our Knowledge of Geographical Zoology," *Nature* 12 (1875), 374–382.

215. Charles Darwin, *On the Origin of Species by Means of Natural Selection, or the Preservation of Favoured Races in the Struggle for Life,* ed. Ernst Mayr (1859; Cambridge, Mass.: Harvard University Press, 1964), chap. 11; Alfred Russel Wallace, *Island Life, or, The Phenomena and Causes of Insular Faunas and Floras* (London: Macmillan, 1880).

216. Frank E. Bedard, *A Text-book of Zoography* (Cambridge: Cambridge University Press, 1895), 183.

217. William Lutley Sclater and Philip Lutley Sclater, *The Geography of Mammals* (London: Kegan Paul, Trench, Trübner, 1899), 1.

218. Leonard Jenyns, "Some Observations on the Common Bat of Pennant: with an Attempt to prove its Identity with the Pipistrelle of French Authors," *Linnean Society Transactions* 16 (1829), 159.

219. Robert Garner, *The Natural History of the County of Stafford, Comprising its Geology, Zoology, Botany and Meteorology. Also Its Antiquities, Topography, Manufactures* (London: John Van Voorst, 1844), 242.

220. Thomas Bell, with the assistance of Robert F. Tomes and Edward Richard Alston, *A History of British Quadrupeds, including the Cetacea* (London: John Van Voorst, 1874), vii.

221. William Shakespeare, *Richard II,* II, 45; F. G. Aflalo, *Types of British Animals* (London: Sands, 1909), 72–73. In fact, then as now, apes and monkeys had many taxonomical relatives in Great Britain.

222. Leonard Jenyns, *A Manual of British Vertebrate Animals* (Cambridge: Pitt Press, 1835), xii; Alfred Heneage Cocks, *Guide to the Collection of British Wild Animals Now or Lately Living in the Private Menagerie at Thames Bank, Great Marlow* (n.p., n.d.).

223. Christopher Frost, *A History of British Taxidermy* (Lavenham, Suffolk: Lavenham Press, 1987), 5.

224. Brookes, *New and Accurate System,* I, 316.

225. Thomas Pennant, *British Zoology* (London: Benjamin White, 1768), 114.

226. Pennant, *British Zoology,* 114; Martyn, *New Dictionary,* I, not paginated.

227. Church, *Cabinet of Quadrupeds,* II, not paginated; Bewick, *General History,* 506; Chappelow, "Sentimental Naturalist," 100.

228. *Zoological Sketches; Consisting of Descriptions of One Hundred and Twenty Animals* (London: Society for Promoting Christian Knowledge, 1844), 1.

229. "The Proboscis-Seal, or Sea-Elephant," *Zoological Magazine, or Journal of Natural History* 1 (1833), 145.

230. Bewick, *General History,* 506, 513.

231. Bigland, *Letters on Natural History,* 265.

232. "The Performing Fish," *Illustrated London News,* May 28, 1854.

233. *Guide to the Ipswich Museum,* 16; "Lectures at the Zoological Gardens," *Nature* 11 (1875), 513.

234. Shaw, *General Zoology,* II, 471; Tyson, *Phocaena,* quoted in Ashley Montagu, *Edward Tyson,* 99.

235. John Reinhold Forster, *A Catalogue of the Animals of North America* (London: B. White, 1771), 19.

236. Kerr, *Animal Kingdom,* 355.

237. Bigland, *Letters on Natural History,* vi.

238. Turner, "Essay on Classification," 1952.

239. Desmond and Moore, *Darwin,* 45; Rennie, *Alphabet of Zoology,* 139.

240. John Stuart Mill, *A System of Logic, Ratiocinative and Inductive: Being a Connected View of the Principles of Evidence and the Methods of Scientific Investigation* (New York: Harper and Brothers, 1881), 500.

241. Flower, "Whales, and British and Colonial Whale Fisheries," in *Essays on Museums,* 187.

242. Richard Lydekker, *Royal Natural History* (London: Frederick Warne, 1895), vol. III, sec. V, 1.

2. Flesh Made Word

1. Quoted in [William Jardine], "Proposed Reform of Zoological Nomenclature," *Edinburgh New Philosophical Journal,* new ser., 18 (1863), 266.

2. The "Methodus" was also reprinted in the second to ninth editions of the *Systema Naturae.* Translated in James L. Larson, *Reason and Experience: The Representation of Natural Order in the Work of Carl von Linné* (Berkeley: University of California Press, 1971), 154–156.

3. For a discussion of Linnaeus's development of this theme, especially in the *Fauna Svecica* (1746), see Sten Lindroth, "The Two Faces of Linnaeus," in Tore Frängsmyr, ed., *Linnaeus: The Man and His Work* (Berkeley: University of California Press, 1983), 25. Linnaeus's reference was to *Genesis* 2:18–20, in which, before God made a "help meet" for Adam, he asked Adam to name all the creatures, to make sure that the "help meet" did not already exist among them. Adam thus appropriated at least a shadow of divine authority, because "whatsover Adam called every living creature, that *was* the name thereof." Only after Adam had named all the animals did it become clear that "there was not found a help meet for him." Quoted from the King James version.

4. John Hill, *An History of Animals. Containing Descriptions of the Birds, Beasts, Fishes . . . Including Accounts of the Several Classes of Animalcules . . .* (London: Thomas Osborne, 1752), "Preface," not paginated.

5. Thomas Pennant, *History of Quadrupeds* (London: B. and J. White, 1793), I, iii–v.

6. [John Fleming], "On Systems and Methods in Natural History. By J. E. Bicheno," *Quarterly Review* 41 (1829), 304.

7. Pennant, *History of Quadrupeds,* I, 111, 157.

8. Edward Tyson, *Phocaena, or the Anatomy of a Porpesse,* quoted in M. F. Ashley Montagu, *Edward Tyson, M.D., F.R.S., 1650–1708, and the Rise of Human and Comparative Anatomy in England* (Philadelphia: American Philosophical Society, 1943), 95.

9. Georges Louis Leclerc, Comte de Buffon, *Barr's Buffon: Buffon's Natural History* (London: H. D. Symonds, 1797), VI, 337.

10. L. C. Rookmaaker, *The Zoological Exploration of Southern Africa, 1650–1790* (Rotterdam: A. A. Balkema, 1989), ix, 3.

11. Charles Hamilton Smith, "Observations on Some Animals of America Allied to the Genus Antilope," *Linnean Society of London Transactions* 13 (1819), 36–37.

12. John Hunter, *Essays and Observations on Natural History, Anatomy, Physiology, Psychology, and Geology,* ed. Richard Owen (London: John Van Voorst, 1861), II, 249.

13. W. Burchell, *A List of Quadrupeds brought by Mr. Burchell from Southern Africa, and presented by him to the British Museum on the 30th of September, 1817* (London: A. Spottiswoode, n.d.), 7.

14. Thomas Stamford Raffles, "Descriptive Catalogue of a Zoological Collection, Made on Account of the Honourable East India Company, in the Island of Sumatra and its Vicinity . . . ," *Linnean Society of London Transactions* 13 (1820), 269–279.

15. George Vasey, *A Monograph of the Genus Bos. The Natural History of Bulls, Bisons, and Buffaloes. Exhibiting All the Known Species and the More Remarkable Varieties* (London: John Russell Smith, 1857), 70.

16. W. Bullock, *A Companion to Mr. Bullock's Museum . . .* (London: W. Bullock, 1810), 125–126.

17. *Sketch for a Natural History of the Four Animals Now Exhibiting in the King's Mews . . .* (London: C. Handy, 1817).

18. Poster in Human Freaks 2, John Johnson Collection of Printed Ephemera, Bodleian Library, Oxford; George M. Gould and Walter L. Pyle, *Anomalies and Curiosities of Medicine* (1896; New York: Julian Press, 1956), 445.

19. C. Waterton, "The Nondescript," *Magazine of Natural History* 6 (1833), 381–383; Richard Aldington, *The Strange Life of Charles Waterton, 1782–1865* (New York: Duell, Sloan and Pearce, 1949), 91–97.

20. Nehemiah Grew stated the rationale for such nonsystematic Latin designations in the preface to *Musaeum Regalis Societatis. Or A Catalogue and Description of the Natural and Artificial Rarities Belonging to the Royal Society* (London: W. Rawlins, 1681): "the Names of Things should be always taken from something more observably declarative of their Form, or Nature . . . For so, every Name were a short Definition."

21. William Lawrence, "Introduction," in J. F. Blumenbach, *A Short System of Com-*

parative Anatomy, trans. William Lawrence (London: Longman, Hurst, Rees, and Orme, 1807), xv.

22. The origin and evolution of Linnaeus's nomenclature has been the object of intense and extended scholarly scrutiny. This discussion has drawn on A. J. Cain, "The Post-Linnaean Development of Taxonomy," *Proceedings of the Linnean Society of London* 170 (1957–58), 234–236; W. T. Stearn, "The Background of Linnaeus's Contributions to the Nomenclature and Methods of Systematic Biology," *Systematic Zoology* 8 (1959), 4–22; John Lewis Heller, "The Early History of Binomial Nomenclature," *Huntia* 1 (1964), 33–70; John Lewis Heller, *Studies in Linnaean Method and Nomenclature* (Frankfurt and New York: Peter Lang, 1983), esp. chap. 9; Jerry Stannard, "Linnaeus, Nomenclator Historicusque Neoclassicus," in John Weinstock, ed., *Contemporary Perspectives on Linnaeus* (Lanham, Md.: University Press of America, 1985), 17–35; and Frans S. Stafleu, *Linnaeus and the Linnaeans: The Spreading of Their Ideas in Systematic Botany, 1735–1789* (Utrecht: Oosthoek, 1971), 103–110, 337.

23. George Shaw, *Museum Leverianum, Containing Select Specimens from the Museum of the late Sir Ashton Lever* (London: James Parkinson, 1792–1796), I, 38.

24. Author's copy. Such emendations were especially frequent for animals like the kangaroo, which attracted zoological attention on other grounds.

25. *Zoological Keepsake; or Zoology, and the Garden and Museum of the Zoological Society, for the Year 1830* (London: Marsh and Miller, 1830), 152.

26. Henry O. Forbes, *A Hand-book to the Primates* (London: Edward Lloyd, 1896–1897), I, vii.

27. Adrian Desmond and James Moore, *Darwin* (London: Michael Joseph, 1991), chap. 22.

28. Ernst Mayr, *The Growth of Biological Thought: Diversity, Evolution, and Inheritance* (Cambridge, Mass.: Harvard University Press, 1982), 240–241.

29. William Henry Flower, *The Horse: A Study in Natural History* (London: Kegan Paul, Trench, Trübner, 1891), 71.

30. Thomas Hardwicke, "Description of a New Genus of the Class Mammalia, from the Himalaya Chain . . . ," *Linnean Society of London Transactions* 15 (1821), 163–164.

31. William Jardine, *The Natural History of the Ruminating Animals, Part II, Containing Goats, Sheep, Wild and Domestic Cattle* (Edinburgh: W. H. Lizars, 1836), 189; John F. Eisenberg, *The Mammalian Radiations: An Analysis of Trends in Evolution, Adaptation, and Behavior* (Chicago: University of Chicago Press, 1981), 205.

32. For discussions of the development of what has become the "International Code of Zoological Nomenclature" in the nineteenth century and after, see David Heppell, "The Evolution of the Code of Zoological Nomenclature," in Alwyne Wheeler and James H. Price, eds., *History in the Service of Systematics* (London: Society for the Bibliography of Natural History, 1981), 135–141; E. G. Linsley and R. L. Usinger, "Linnaeus and the Development of the International Code of

Zoological Nomenclature," *Systematic Zoology* 8 (1959), 39–47; and Antonello La Vergata, "Au Nom de l'Espèce. Classification et Nomenclature au XIXe Siècle," in Scott Atran et al., *Histoire du Concept de l'Espèce dans les Sciences de la Vie* (Paris: Fondation Singer-Polignac, 1987), 193–225.

33. Solitarius, "Remarks upon Zoological Nomenclature and Systems of Classification," *Field Naturalist* 1 (1833), 523.

34. Nomenclature Papers, Hugh E. Strickland Collection, Cambridge University Museum, Scrapbook I, not paginated. Original committee members included Darwin, J. S. Henslow, L. Jenyns, W. Ogilby, J. Phillips, J. Richardson, H. E. Strickland, and J. O. Westwood. W. J. Broderip, Owen, W. E. Shuckard, G. R. Waterhouse, and W. Yarrell joined later.

35. *Proposed Plan for Rendering the Nomenclature of Zoology Uniform and Permanent* (London: Richard and John E. Taylor, 1841); *Proposed Report of the Committee on Zoological Nomenclature* (London, 1842); Hugh Strickland, "Report of a Committee appointed 'to consider of the rules by which the Nomenclature of Zoology may be established on a uniform and permanent basis,'" *Report of the British Association for the Advancement of Science for 1842*, 105–121. For a detailed account of this nomenclatural initiative in the context of Victorian debate about the concept of species, see Gordon McOuat, "Species, Rules and Meaning: The Politics of Language and the Ends of Definition in Nineteenth-Century Natural History," *Studies in History and Philosophy of Science* 27 (1996), 473–520.

36. Jardine, "Proposed Reform," 260–261; Heppell, "Evolution of the Code," 136–137.

37. Alfred Russel Wallace, "Zoological Nomenclature," *Nature* 9 (1874), 258–259.

38. William Henry Flower, "A Century's Progress in Zoological Knowledge," *Essays on Museums and Other Subjects Concerned with Natural History* (1898; rpt., New York: Books for Libraries Press, 1972), 167.

39. Mrs. Hugh E. Strickland and W. Jardine, eds., *Ornithological Synonyms by the late Hugh Edwin Strickland* (London: John Van Voorst, 1855), I, iii.

40. Alfred Newton, "More Moot Points in Ornithological Nomenclature," *Annals and Magazine of Natural History* 4 (1879), 419–420.

41. *Proposed Plan*, 2, 3, 9.

42. W. J. Broderip to Hugh Strickland, May 5, 1842, in Nomenclature Papers, Strickland Collection, Scrapbook II.

43. *Proposed Plan*, 1.

44. Thomas Stamford Raffles (Communicated by Everard Home), "Descriptive Catalogue of a Zoological Collection, made on account of the East India Company, in the Island of Sumatra and its Vicinity," *Linnean Society of London Transactions* 13 (1820–1821), 239–240.

45. Janet Browne, *Charles Darwin: Voyaging* (New York: Knopf, 1995), 360.

46. "Notes," *Nature* 10 (1874), 453.

47. Strickland, "Report of a Committee," 106–107.

48. Lawrence, "Introduction," xvi; J. E. Bicheno, "On Systems and Methods in Natural History," *Linnean Society of London Transactions* 15 (1827), 494.

49. John Richardson to Hugh Strickland, March 1, 1842, Nomenclature Papers, Strickland Collection, Scrapbook I.

50. Heppell, "Evolution of the Code," 136–137; Richard V. Melville, *Towards Stability in the Names of Animals: A History of the International Commission on Zoological Nomenclature, 1895–1995* (London: International Trust for Zoological Nomenclature, 1995), 7–20.

51. *Proposed Plan,* 2.

52. William Cornwallis Harris, *Narration of an Expedition into Southern Africa, During the Years 1836 and 1837* (Bombay: American Mission Press, 1838), 256.

53. Richard Lydekker, *Sir William Flower* (London: J. N. Dent, 1906), 97.

54. For example, Lionel Walter Rothschild, an aggressive collector who sent his proxies to gather the zoological spoils of the British Empire in the late nineteenth and early twentieth centuries, was commemorated in the name of 58 species or subspecies of birds, 18 of mammals, three of fish, two of reptiles and amphibia, 153 of insects, three of arachnids, one of millipedes, and one of nematodes. Miriam Rothschild, *Dear Lord Rothschild: Birds, Butterflies and History* (Philadelphia: Balaban, 1983), 364. For further examples, see Harriet Ritvo, "The Power of the Word: Scientific Nomenclature and the Spread of Empire," *Victorian Newsletter* (Spring 1990), 5–8.

55. "The Zoological Record," *Nature* 18 (1878), 485.

56. Solitarius, "Faults in Zoological Nomenclature," *Field Naturalist* 1 (1833), 461.

57. Charles Darwin to H. E. Strickland, January 29, 1849, in Nomenclature Papers, Strickland Collection, Scrapbook I; *Correspondence of Charles Darwin,* ed. F. Burkhardt and Sydney Smith (Cambridge: Cambridge University Press, 1988), IV, 210; Solitarius, "Remarks upon Zoological Nomenclature," 522.

58. For extended discussions of the role of class within the early Victorian scientific community, see Martin Rudwick, *The Great Devonian Controversy* (Chicago: University of Chicago Press, 1986), and Adrian Desmond, *The Politics of Evolution: Morphology, Medicine and Reform in Radical London* (Chicago: University of Chicago Press, 1990). For an analysis of social influence on French scientific nomenclature, see Yves Delaporte, "*Sublaevigatus* ou *Subloevigatus?* Les usages sociaux de la nomenclature chez les entomologistes," in Jacques Hainard and Roland Kaehr, eds., *Des Animaux et des Hommes* (Neuchâtel, Switzerland: Musée d'Ethnographie, 1987), 187–212.

59. *Proposed Plan,* 11–14.

60. Charles Lyell, *Principles of Geology* (1830–1833; rpt., Chicago: University of Chicago Press, 1990–1991), III, 53.

61. *Proposed Plan,* 5.

62. *Proposed Plan,* 11–14.

63. Thomas Hawkins, *The Book of the Great Sea Dragons, Ichthyosauri and Plesiosauri* . . . (London: William Pickering, 1840), 9.

64. Francis P. Pascoe, *Zoological Classification: A Handy Book of Reference, with Tables of the Subkingdoms, Classes, Orders, &c of the Animal Kingdom, Their Charac-*

ters, and Lists of the Families and Principal Genera (London: John Van Voorst, 1880), vi.

65. James Edward Smith, *An Introduction to Physiological and Systematical Botany* (London: Longman, Hurst, Rees, Orme, and Brown, 1819), vi.

66. John Ruskin, *Love's Meinie. Lectures on Greek and English Birds* (1881), vol. XXV, *The Works of John Ruskin,* ed. E. T. Cook and Alexander Wedderburn (London: George Allen, 1906), 21–22, 14.

67. Edward Lear, *The Complete Nonsense Book,* ed. Lady Strachey (New York: Dodd, Mead, 1934), 197–219.

68. Charles Kingsley, *The Water Babies* (1863; London: Penguin, 1995), 157–158.

69. *Punch* 71 (1876), 101; *Punch* 61 (1871), 194.

70. Elizabeth Gaskell, *Mary Barton: A Tale of Manchester Life* (1848; London: J. M. Dent, 1911), 144. The form of the Latin parody may be unrealistically precise.

71. *Proposed Plan,* 3.

72. "Animals by their Proper Names," *Oriental Sporting Magazine,* no. 20 (1833), 411.

73. Richard Lydekker, *Horns and Hoofs, or Chapters on Hoofed Animals* (London: Horace Cox, 1893), 194, 2, 32, 149.

74. James Watson Roberts, "Of the Degeneration of Animals," Society for Investigating Natural History, vol. 4 (1785–1796), 96. Minute books in Special Collections, Edinburgh University Library.

75. Harriet Ritvo, *The Animal Estate: The English and Other Creatures in the Victorian Age* (Cambridge, Mass.: Harvard University Press, 1987), 214.

76. E. T. Bennett, *The Gardens and Menagerie of the Zoological Society Delineated* (Chiswick: Charles Tilt, 1831), vii.

77. "Report to the Vice-Chancellor on the Removal of the Collection of Comparative Anatomy. October 19, 1865," Item 69 in the Professor of Anatomy Collection, Cambridge University Archives; "Returns Relating to the National Collections of Works of Art &c," *Parliamentary Papers* 13 (1857), 53. The term *trivial* has traditionally been used by naturalists in two, rather contradictory senses: to indicate the specific, as opposed to the generic component of the binomial, and to indicate the vernacular as opposed to scientific term for an organism.

78. Banks's crusade has been described in detail in Harold Burnell Carter, *His Majesty's Spanish Flock: Sir Joseph Banks and the Merinos of George III of England* (Sydney: Angus and Robertson, 1964).

79. See James A. Secord, "Darwin and the Breeders: A Social History," in David Kohn, ed., *The Darwinian Heritage* (Princeton: Princeton University Press, 1985), 519–542, and, generally, Charles Darwin, *The Variation of Animals and Plants under Domestication* (New York: Orange Judd, 1868).

80. Everett Millais, *The Theory and Practice of Rational Breeding* (London: The "Fancier's Gazette," 1869), vii–ix.

81. J. C. Ewart, *The Pennycuik Experiments* (London: Adam and Charles Black, 1899), 63.

82. *Punch* 54 (February 29, 1868), 91.

83. Scott dedicated several stanzas to the cattle in "Cadyow Castle," published in *Minstrelsy of the Scottish Border* (1802); Landseer exhibited *Death of the Wild Bull* at the Royal Academy show of 1836 and *The Wild Cattle of Chillingham* at the Royal Academy show of 1867.

84. For a more extensive discussion of the cattle in relation to eighteenth- and nineteenth-century British culture, see Harriet Ritvo, "Race, Breed, and Myths of Origin: Chillingham Cattle as Ancient Britons," *Representations* 39 (Summer 1992), 1–22.

85. Juliet Clutton-Brock, *A Natural History of Domesticated Mammals* (London: British Museum (Natural History), Cambridge: Cambridge University Press, 1987), 63.

86. Thomas Bewick, *A General History of Quadrupeds* (Newcastle: T. Bewick, 1824), 38.

87. John Storer, *The Wild White Cattle of Great Britain. An Account of their Origin, History, and Present State*, ed. John Storer, Jr. (London: Cassell, 1879), xv.

88. James Edmund Harting, *British Animals Extinct within Historic Times, with some account of British Wild White Cattle* (London: Trübner, 1880), 213.

89. Jan Tavinor, "A Chapter in the History of the 'Chartleys,'" *Ark* 18 (1991), 379.

90. A. G. Bradley, *The Romance of Northumberland* (1908; rpt., London: Methuen, n.d.), 89.

91. Robert Chambers, *Vestiges of the Natural History of Creation* (New York: Wiley and Putnam, 1846), 236.

92. See, for example, Boyd Dawkins, "The Chartley White Cattle," *North Staffordshire Field Club Annual Report and Transactions* 33 (1898–99), 49.

93. Modern zoological consensus derives all European and Asian domestic cattle from the aurochs. The other species identified by Victorian paleontologists were early domestic cattle. Clutton-Brock, *Natural History of Domesticated Mammals*, 63–65.

94. Jacob Wilson, "The Chillingham Wild Cattle," *Land Magazine* (January 1899), 19–20.

95. Harting, *British Animals Extinct within Historic Times*, 213–214, 216–217.

96. H. T. Meunell and V. R. Perkins, "Wild Cattle of Chillingham," *Transactions of the Tyneside Naturalists Field Club* 6 (1863–64), 142–143.

97. George Henry Andrews, *Modern Husbandry: A Practical and Scientific Treatise on Agriculture* . . . (London: Nathaniel Cooke, 1853), 150.

98. Pennant, *History of Quadrupeds*, I, 17.

99. For example, James Aikman, *A Natural History of Beasts, Birds, and Fishes: or Stories of Animated Nature* (Edinburgh: Thomas Nelson, 1846), 37.

100. William Swainson, *On the Natural History and Classification of Quadrupeds* (London: Longman, 1835), 235; disparagingly discussed in Vasey, *Monograph of the Genus Bos*, 178–179.

101. Thomas Bell, assisted by Robert F. Tomes, *A History of British Quadrupeds, including the Cetacea* (London: John Van Voorst, 1874), 368; Vasey, *Monograph of the Genus Bos*, 140.

102. "Wild Cattle of Great Britain," *Penny Magazine* 7 (1838), 443.

103. *Sussex Herd Book, Containing the Names of the Breeders, the Age, and the Pedigrees of the Sussex Cattle,* II (1885), "Preface," not paginated.

104. F. A. Manning, "The History and Growth of the Kennel Club," *Dog Owners' Annual* (1890), 138.

105. For a discussion of the construction of domestic animal breeds, by Robert Bakewell and others, see Harriet Ritvo, "Possessing Mother Nature: Genetic Capital in Eighteenth-Century Britain," in John Brewer and Susan Staves, eds., *Early Modern Conceptions of Property* (London: Routledge, 1995), 413–426. John Borneman has offered a synchronic account of a similar process in "Race, Ethnicity, Species, Breed: Totemism and Horse-Breed Classification in America," *Comparative Studies in Society and History* 30 (1988), 25–51.

106. Leonard Bull, *History of the Smithfield Club from 1798 to 1925* (London: Smithfield Club, 1926), 261; George Coates, *The General Short-Horned Herd-Book: Containing the Pedigrees of Short-Horned Bulls, Cows, &c. of the Imported Durham Breed* (Otley: W. Walker, 1822).

107. Stephen J. G. Hall and Juliet Clutton-Brock, *Two Hundred Years of British Farm Livestock* (London: British Museum (Natural History), 1989), 49.

108. David Low, *On the Domesticated Animals of the British Islands* (London: Longman, Brown, Green and Longmans, 1845), 381–382.

109. C. M. Prior, *The History of the Racing Calendar and the Stud-Book, from their Inception in the Eighteenth Century, with Observations on Some of the Occurrences Noted Therein* (London: Sporting Life, 1926), 15. See Nicholas Russell, *Like Engend'ring Like: Heredity and Animal Breeding in Early Modern England* (Cambridge: Cambridge University Press, 1986), chap. 5, for a fuller account of eighteenth-century horse-breeding.

110. James Long, *The Book of the Pig: Its Selection, Breeding, Feeding, and Management* (London: L. Upcott Gill, 1880), 316, 124.

111. Robert Knox, *The Races of Men: A Philosophical Enquiry into the Influence of Race over the Destinies of Nations* (London: Henry Renshaw, 1862), 2.

112. Judith Neville Lytton, *Thoroughbred Racing Stock and Its Ancestors. The Authentic Origin of Pure Blood* . . . (London: George Allen and Unwin, 1938), 64–65.

113. James Anderson, *Essays Relating to Agriculture and Rural Affairs* (Edinburgh: William Creech, 1777), II, 138–139.

114. Charles Hassall, *General View of the Agriculture of the County of Carmarthen* (London: W. Smith, 1794), 35, 37.

115. William Bingley, *Animal Biography; or, Authentic Anecdotes of the Lives, Manners and Economy of the Animal Creation* (London: Richard Phillips, 1804), I, 202.

116. This iteration occurred in Thomas Pennant, *British Zoology* (London, 1768), I, 50. A Victorian facsimile of *Of Englishe Dogges* was published in 1880.

117. Everett Millais, "A Colouring Tonic for Dogs," *Dog Owners' Annual for 1892* (London: Dean and Son, 1892), 86.

118. S. O. Beeton, *Beeton's Book of Home Pets: Showing How to Rear and Manage, in*

Sickness and in Health, Birds, Poultry, Pigeons, Rabbits, Guinea Pigs, Dogs, Cats, Squirrels, Fancy Mice, Tortoises, Bees, Silkworms, Ponies, Donkeys, Goats, Inhabitants of the Aquarium, Etc. (London: Ward, Lock and Tyler, n.d. [late 1800s]), 551 ff.

119. B. Waterhouse Hawkins, *The Artistic Anatomy of the Dog and the Deer* (London: Winsor and Newton, 1876), 9.

120. See, for example, H. D. Richardson, *The Dog: Its Origin, Natural History, and Varieties* (London: William S. Orr, 1853), 40.

121. Peter Lund Simmonds, *Animal Products. Their Preparation, Commercial Uses, and Value* (New York: Scribner, Welford, and Armstrong, 1877), 237; Edward William Jaquet, *The Kennel Club: A History and Record of its Work* (London: Kennel Gazette, 1905), 19–20.

122. Charles Hamilton Smith, *Dogs . . . Part II. (Domestic Dogs, Foxes, and Hyaenas)* (Edinburgh: W. H. Lizars, 1840); Edward Griffith, *General and Particular Descriptions of the Vertebrated Animals Arranged Conformably to Modern Discoveries and Improvements in Zoology. Order Carnivora* (London: Baldwin, Craddock, and Joy, 1821), 204–218.

123. John Fleming, *A History of British Animals, Exhibiting the Descriptive Characters and Systematical Arrangement of the Genera and Species of Quadrupeds* (Edinburgh: Bell and Bradfute, 1828), 11.

124. *Guide to the Gardens of the Zoological Society, March 1829* (London: Richard Taylor, 1829), 5.

125. Vero Shaw, *The Illustrated Book of the Dog* (London: Cassell, Petter, Galpin, 1881), 2–5.

126. On the social dimension of nineteenth-century British livestock breeding and pet fancying, see Ritvo, *Animal Estate*, chaps. 2 and 3.

127. J. Mortimer, *The Whole Art of Husbandry: Or, The Way of Managing and Improving of LAND* (London: R. Robinson, 1721), I, 227.

128. W. C. L. Martin, "The Ox," in *The Farmer's Library. Animal Economy* (London: Charles Knight, 1849), I, 56.

129. A. Coventry, *Remarks on Live Stock and Relative Subjects* (Edinburgh: 1806), 39.

130. John Wilkinson, *Remarks on the Improvement of Cattle, &c. in a Letter to Sir John Saunders Sebright, Bart.* (Nottingham: H. Barnet, 1820), 4–5; J. Cossar Ewart, "The Principles of Breeding and the Origin of Domesticated Breeds of Animals," in *Twenty-Seventh Annual Report of the Bureau of Animal Industry [Department of Agriculture] for the Year 1910* (Washington, D.C.: Government Printing Office, 1912), 152.

131. Henry F. Euren, *The Norfolk and Suffolk Red Polled Herd Book*, I (1874), 7.

132. James Sinclair, ed., *History of Shorthorn Cattle* (London: Vinton, 1907), 97.

133. Robert Bruce, *Fifty Years Among Shorthorns, with over 300 Pen Pictures of Notable Sires* (London: Vinton, 1907), 195.

134. James MacDonald and James Sinclair, *History of Polled Aberdeen or Angus Cattle, Giving an Account of the Origin, Improvement, and Characteristics of the Breed*

(Edinburgh: William Blackwood, 1882), chap. 17; "Breeder's Directory," appended to Robert William Ashburner, *The Shorthorn Herds of England, 1885–6–7* (Warwick: Henry T. Cooke, 1888); Bruce, *Fifty Years Among Shorthorns,* 134–152.

135. William Swainson, *A Preliminary Discourse on the Study of Natural History* (London: Longman, Rees, Orme, Brown, Green, and Longman, 1834), 232.

136. Everett Millais, *Two Problems of Reproduction* (Manchester: "Our Dogs" Publishing Company, 1895), 4.

137. Darwin, *Variation of Animals and Plants under Domestication,* II, 297–298.

138. [Whitwell Elwin], "Blaze's *History of the Dog,*" *Quarterly Review* 72 (1843), 488; W. Holt Beever, *An Alphabetical Arrangement of the Leading Shorthorn Tribes with Notes for the Use of Breeders* (London: J. Thornton, n.d. [ca. 1885]), v.

3. Barring the Cross

1. On pre-Darwinian evolutionary debate in Britain, see Adrian Desmond, *The Politics of Evolution: Morphology, Medicine, and Reform in Radical London* (Chicago: University of Chicago Press, 1989).

2. Charles Darwin, *On the Origin of Species: A Facsimile of the First Edition,* ed. Ernst Mayr (1859; Cambridge, Mass.: Harvard University Press, 1964), 469–470.

3. Darwin, *Origin,* 484. On the taxonomically conservative aspects of Darwin's theory, see Harriet Ritvo, "Classification and Continuity in *The Origin of Species,*" in *Charles Darwin's The Origin of Species: New Interdisciplinary Essays,* ed. David Amigoni and Jeff Wallace (Manchester: Manchester University Press, 1995), 47–67.

4. Charles Lyell, *The Geological Evidences of the Antiquity of Man, with remarks on theories of the Origin of Species by Variation* (London: John Murray, 1863), 388.

5. George Henry Lewes, *Studies in Animal Life* (London: Smith, Elder, 1862), 128–129.

6. William Bernhard Tegetmeier and C. L. Sutherland, *Horses, Asses, Zebras, Mules, and Mule Breeding* (London: Horace Cox, 1895), 49.

7. F. G. Aflalo, *Types of British Animals* (London: Sands, 1909), 4.

8. For recent historical accounts of this debate, see Peter F. Stevens, "Species: Historical Perspectives," in *Keywords in Evolutionary Biology,* ed. Evelyn Fox Keller and Elisabeth A. Lloyd (Cambridge, Mass.: Harvard University Press, 1992), 302–311; Scott Atran, *Cognitive Foundations of Natural History: Towards an Anthropology of Science* (Cambridge: Cambridge University Press, 1990), chap. 6; and Ernst Mayr, *The Growth of Biological Thought: Diversity, Evolution, and Inheritance* (Cambridge, Mass.: Harvard University Press, 1982), chap. 6.

9. Henri Daudin, *De Linné à Lamarck: Méthodes de la Classification et Idée de Série en Botanique et en Zoologie* (Paris: Librairie Félix Alcan, 1926), 229.

10. Darwin, *Origin,* 177.

11. Thomas Boreman, *A Description of Some Curious and Uncommon Creatures, Omitted in the Description of Three Hundred Animals . . .* (London: Richard Ware and Thomas Boreman, 1739), 29.

12. "White-Footed Antelope," *The Naturalist's Pocket Magazine* 2 (1799), not paginated.

13. David Low, *The Breeds of the Domestic Animals of the British Islands* (London: Longman, Orme, Brown, Green and Longmans, 1842), vol. II, *Goats*, 1.

14. "Lectures at the Zoological Gardens. IV. May 13—Mr. Garrod on Antelopes and their Allies," *Nature* 12 (May 27, 1875), 68.

15. George Shaw, *General Zoology; or Systematic Natural History* (London: G. Kearsley, 1800), II, 205; II, 239–240, 244.

16. Shaw, *General Zoology*, II, 127.

17. Shaw, *General Zoology*, II, 127; Thomas Stamford Raffles, "Descriptive Catalogue of a Zoological Collection, made on account of the Honourable East India Company, in the Island of Sumatra and its Vicinity . . . ," *Linnean Society of London Transactions* 13 (Read December 5, 1820), 259.

18. Robert Knox, *The Races of Men: A Philosophical Enquiry into the Influence of Race over the Destinies of Nations* (London: Henry Renshaw, 1862), 65.

19. Edward Griffith, *General and Particular Descriptions of the Vertebrated Animals, Arranged Conformably to the Modern Discoveries and Improvements in Zoology. Order Carnivora* (London: Baldwin, Cradock, and Joy, 1821), 109–110.

20. Thomas Bewick, *A General History of Quadrupeds* (Newcastle upon Tyne: T. Bewick, 1824), 16.

21. John Fleming, *The Philosophy of Zoology; or A General View of the Structure, Functions, and Classifications of Animals* (Edinburgh: Archibald Constable, 1822), I, 429.

22. Thomas Pennant, *History of Quadrupeds* (London: B. and J. White, 1793), I, 29.

23. Georges Louis Leclerc, Comte de Buffon, *Barr's Buffon. Buffon's Natural History* (London: H. D. Symonds, 1797), VI, 146–147.

24. John Sherer, *Rural Life Described and Illustrated* (London: London Printing and Publishing Company, 1868), 370.

25. James Watson Roberts, "Of the Degeneration of Animals," Society for Investigating Natural History/Natural History Society, *Papers Delivered* 4 (1785–86), 107. Manuscript in Special Collections, Edinburgh University Library.

26. John Bigland, *Letters on Natural History: Exhibiting a View of the Power, Wisdom, and Goodness of the Deity* . . . (London: Longman, Hurst, Rees and Orme, 1806), 230; John French Burke, *Farming for Ladies; Or, a Guide to the Poultry-Yard, the Dairy and Piggery* (London: John Murray, 1844), 324.

27. John Fry, "On Factitious or Mule-Bred Animals," *Hippiatrist and Veterinary Journal* 3 (1830), 113, 115.

28. Juliet Clutton-Brock, *Horse Power: A History of the Horse and the Donkey in Human Societies* (Cambridge, Mass.: Harvard University Press, 1992), 43–44.

29. Philip Gosse, *An Introduction to Zoology* (London: Society for Promoting Christian Knowledge, 1844), I, xv.

30. Tegetmeier and Sutherland, *Horses*, 80. A mule is the offspring of a female horse and a male donkey; a hinny is the offspring of a female donkey and a male horse.

31. George Garrard, *A Description of the Different Varieties of Oxen, Common in the British Isles* (London: J. Smeeton, 1800), not paginated.

32. Thomas Eyton, "Some Remarks upon the Theory of Hybridity," *Magazine of Natural History* 1 (1837), 358; John Jones, *Medical, Philosophical, and Vulgar Errors, of Various Kinds, Considered and Refuted* (London: T. Cadell Jun. and W. Davies, 1797), 101.

33. Paul Broca, *On the Phenomena of Hybridity in the Genus Homo,* ed. C. Carter Blake (London: Longman, Green, Longman, and Roberts/Anthropological Society, 1864), x; Jean-Louis Fischer, "L'Hybridologie et la Zootaxie du Siècle des Lumières à L'Origine des Espèces," *Revue de Synthèse* 102 (1981), 64–66.

34. Charles Lyell, *Principles of Geology,* ed. Martin Rudwick (1832; Chicago: University of Chicago Press, 1991), II, 49.

35. *The Animal Kingdom, or Zoological System of the Celebrated Sir Charles Linnaeus; Class I, Mammalia . . . ,* trans. Robert Kerr (London: J. Murray, 1792), 346; "Mongooz," *The Naturalist's Pocket Magazine* 2 (1799), not paginated.

36. E. P. Evans, *The Criminal Prosecution and Capital Punishment of Animals: The Lost History of Europe's Animal Trials* (1906; London: Faber and Faber, 1987), 147–153.

37. Charles White, *An Account of the Regular Gradation in Man, and in Different Animals and Vegetables; and from the Former to the Latter* (London: C. Dilly, 1799), 34; Johann Friedrich Blumenbach, *The Anthropological Treatises . . . ,* ed. and trans. Thomas Bendyshe (London: Longman, Green, Longman, Roberts and Green/The Anthropological Society, 1865), 73.

38. Quoted in Dudley Wilson, *Signs and Portents: Monstrous Births from the Middle Ages to the Enlightenment* (Routledge: London, 1993), 56–57.

39. Edward Tyson, *Orang-Outang, sive Homo Sylvestris. Or, the Anatomy of a Pygmie Compared with that of a Monkey, an Ape, and a Man* (London: Thomas Bennet, 1699), 2.

40. White, *Account of the Regular Gradation in Man,* 34.

41. Jan Bondeson and A. E. W. Miles, "Julia Pastrana, the Nondescript: An Example of Congenital Generalized Hypertrichosis Terminalis with Gingival Hyperplasia," *American Journal of Medical Genetics* 47 (1993), 199.

42. Rajendra Mallika, "Exhibit of a Dead Hybrid Monkey," *Proceedings of the Asiatic Society of Bengal* 32 (1863), 455–456.

43. P. L. Sclater, "Notice of Some Hybrid Monkeys," *Proceedings of the Zoological Society of London* (November 5, 1878), 791.

44. A. D. Bartlett, "Notes on Some Young Hybrid Bears," *Proceedings of the Zoological Society of London* (1860), pt. 28, 130.

45. H. Scherren, "Some Notes on Hybrid Bears," *Proceedings of the Zoological Society of London* (1907), 433–435; Annie P. Gray, *Mammalian Hybrids: A Check-List with Bibliography* (Slough: Commonwealth Agricultural Bureaux, 1972), 57–58.

46. John LeKeux, *Illustrations of Natural History: Embracing a Series of Engravings and Descriptive Accounts of the Most Interesting and Popular Genera and Species of the Animal World. I. Quadrupeds* (London: Longman, 1829–1830), 151; J. Jenner Weir, "Hybrid between Goat and Sheep," *Zoologist,* 3d ser., 12 (1888), 104. Twentieth-century science suggests that although "hybridization has been reported" between the

domestic sheep and the domestic goat, "the two species do not readily interbreed." Gray, *Mammalian Hybrids*, 130–131.

47. See, for example, T——n, "Cross between a Wolf and Hound," *Notes and Queries*, no. 64 (January 18, 1851), 39, and James Burton, "Collectanea AEgyptica. Natural History," 6–7, British Library Add. Ms. 25,666. St. George Mivart, *Dogs, Jackals, Wolves, and Foxes: A Monograph on the Canidae* (London: R. H. Porter, 1890), 22.

48. See, for example, Tegetmeier and Sutherland, *Horses*, chap. 10, and Richard Lydekker, *The Horse and Its Relatives* (London: George Allen, 1912), chap. 10.

49. P. L. Sclater, "On the Zebra-and-Pony Hybrid," *Proceedings of the Zoological Society of London* 1 (1903), 1–2.

50. Vero Shaw, *The Encyclopedia of the Stable: A Complete Manual of the Horse, Its Breeds, Anatomy, Physiology, Diseases, Breeding, Training and Management* (London: George Routledge, 1909), 363; Richard Lydekker, *Mostly Mammals: Zoological Essays* (New York: Dodd Mead, 1903), 43.

51. Richard Lydekker, *Horns and Hoofs, or Chapters on Hoofed Animals* (London: Horace Cox, 1893), 21–29.

52. Scherren, "Some Notes on Hybrid Bears," 432; Charles Hamilton Smith, *The Natural History of Dogs. Volume II. Canidae or Genus Canis of Authors. Including Also the Genera Hyaena and Proteles* (Edinburgh: W. H. Lizars, 1840), 98.

53. Edward Griffith, *General and Particular Descriptions of the Vertebrated Animals . . . Order Carnivora*, 58; George R. Jesse, *Researches into the History of the British Dog, from Ancient Laws, Charters and Historical Records* (London: Robert Hardwick, 1866), I, 337, 338; "Curious Ram," *Annals of Sporting and Fancy Gazette* 1 (1822), 198; "Remarkable Hybrid," *Farrier and Naturalist* 1 (1828), 86.

54. Charles Gould, *Mythical Monsters* (London: W. H. Allen, 1886), 364.

55. Richard Thursfield, "Account of a Hybrid between a Hare and a Rabbit," *Proceedings of the Zoological Society of London* (1830–1831), pt. I, 66.

56. "A Hybrid, Bred by a Hare and a Rabbit," *Veterinary Examiner, or Monthly Record, of Physiology, Pathology, Agriculture and Natural History* 1 (1833), 88.

57. See, for example, Lewes, *Studies in Animal Life*, 161–162. Charles Darwin accepted reports of these hybrids "with hesitation," and was subsequently relieved when they were discredited; *The Variation of Animals and Plants under Domestication* (New York: Orange Judd, 1868), II, 124. Jean-Louis Fischer has discussed the role of the leporides in French debate in "Espèce et Hybrides: A Propos de Léporides," in Scott Atran et al., eds., *Histoire du Concept de l'Espèce dans les Sciences de la Vie* (Paris: Fondation Singer-Polignac, 1987), 252–268.

58. Charles Darwin, *The Correspondence of Charles Darwin*, ed. Frederick Burkhardt et al. (Cambridge: Cambridge University Press, 1986), II, 447.

59. "Portrait of a Cross of the Dog and the Fox," *Annals of Sporting and Fancy Gazette* 6 (1824), 203.

60. Stonehenge [John Henry Walsh], *The Dog in Health and Disease* (London: Longman, Green, Longman and Roberts, 1859), 166, 168.

61. *A Guide to the Ipswich Museum* (Ipswich: Ipswich Museum/J. Haddock, 1871), 14.

62. W. W. Thompson, "Dog and Fox Cross," *Kennel Gazette* 2 (1881), 174.

63. Gray, *Mammalian Hybrids*, 95.

64. Pennant, *History of Quadrupeds*, II, 321.

65. H. D. Richardson, "Ass," in John C. Morton, ed., *A Cyclopedia of Agriculture, Practical and Scientific* (Glasgow: Blackie and Sons, 1855), I, 164–165.

66. Shaw, *Encyclopedia of the Stable*, 218; J. C. Ewart, *The Penycuik Experiments* (London: Adam and Charles Black, 1899), lxxxix.

67. Bewick, *General History of Quadrupeds*, 313–314; Mivart, *Dogs*, viii.

68. See Richard Lydekker, *The Ox and Its Kindred* (London: Methuen, 1912), chap. 10, for a summary of nineteenth-century bovine hybridization.

69. A. D. Bartlett, "On Some Hybrid Bovine Animals . . . ," *Proceedings of the Zoological Society of London* (1884), 400–401.

70. Richard Lydekker, *A Handbook to the Carnivora. Pt. I. Cats, Civets, and Mungooses* (London: Edward Lloyd, 1896), 45–48.

71. From posters in the Cambridgeshire Collection, Cambridge Central Library.

72. *The Picture of Liverpool or Stranger's Guide* (Liverpool, 1833), 216; Harriet Ritvo, *The Animal Estate: The English and Other Creatures in the Victorian Age* (Cambridge, Mass.: Harvard University Press, 1987), 223, 235.

73. A. E. Shipley, *"J.": A Memoir of John Willis Clark, Registrar of the University of Cambridge and Sometime Fellow of Trinity College* (London: Smith, Elder, 1913), 263; J. B. Nelson, "Notice of the Lion-tiger which died in Cambridge, March 1833," Item 19a in Cambridge University Museum of Zoology, History Index I (Catalogue of Acquisitions 1819–1870).

74. "Report on the condition of the anatomical museum for 1864," Item 58 in Cambridge University Register 39.13 (Professor of Anatomy), Cambridge University Archives; Ernst Hartert, *Guide to the Hon. Walter Rothschild's Museum at Tring* (Tring, 1898), 14; R. Lydekker, "Domesticated Animals, Hybrids and Abnormalities," in *The History of the Collections Contained in the Natural History Departments of the British Museum* (London: Trustees of the British Museum, 1906), II, 68.

75. Abraham D. Bartlett, *Wild Animals in Captivity, Being an Account of the Habits, Food, Management and Treatment of the Beasts and Birds at the 'Zoo,' with Reminiscences and Anecdotes*, ed. Edward Bartlett (London: Chapman and Hall, 1899), 216.

76. Scherren, "Some Notes on Hybrid Bears," 434.

77. Gray, *Mammalian Hybrids*, 32, 33, 43, 44, 94–113, 148, 152, 166, 170.

78. Quoted in P. Chalmers Mitchell, *Centenary History of the Zoological Society of London* (London: Zoological Society of London, 1929), 93–94.

79. *Report on the Farm of the Zoological Society at Kingston Hill, March 1832* (London: Richard Taylor, 1832), 7, 9.

80. Bristol, Clifton and West of England Zoological Society, Subcommittee for Zoology, Minutes, 1836–1850 (May 28, 1836), 3, in the zoo's archives.

81. S. J. Woolfall, "History of the 13th Earl of Derby's Menagerie and Aviary at Knowsley Hall, Liverpool (1806–1851)," *Archives of Natural History* 17 (1990), 6; T., "A Visit to Knowsley Hall, in Lancashire, the Seat of the Earl of Derby," *Annals of Sporting and Fancy Gazette* 6 (1824), 224.

82. Society for the Acclimatisation of Animals, Birds, Fishes, Insects and Vegetables within the United Kingdom, *Second Annual Report* (1862), 7.

83. On the success of attempts at acclimatization, see Ritvo, *Animal Estate*, 237–240, and Warwick Anderson, "Climates of Opinion: Acclimatization in Nineteenth-Century France and England," *Victorian Studies* 35 (1992), 147–150.

84. "Acclimatization and Preservation of Animals," *National Review* 17 (1863), 168.

85. J. Cossar Ewart, *Guide to the Zebra Hybrids etc. on Exhibition at the Royal Agricultural Society's Show, York, Together with a Description of Zebras, Hybrids, Telegony, etc.* (Edinburgh: T. and A. Constable, 1900).

86. *Times,* March 13, 1899; *Morning Post,* April 5, 1899.

87. Ewart, *Guide to the Zebra Hybrids,* 1–8.

88. Arthur Shipley, "Zebras, Horses, and Hybrids," *Quarterly Review* 190 (1899), 420.

89. Ewart, *Guide to the Zebra Hybrids,* 41–46.

90. Ewart, *Guide to the Zebra Hybrids,* 2.

91. Ewart, *Guide to the Zebra Hybrids,* 30.

92. See Darwin, *Origin,* chap. 8; Darwin, *Variation of Animals and Plants under Domestication,* II, chaps. 15, 19.

93. Darwin, *Variation of Animals and Plants under Domestication,* I, 39–40.

94. For representative claims, see William Ridgeway, *The Origin and Influence of the Thoroughbred Horse* (1905; New York: Benjamin Blom, 1972), 5; Jacob Wilson, "The Chillingham Wild Cattle," *Land Magazine* (January 1899), 19; Richard Lydekker, ed., *The Royal Natural History* (London: Frederick Warne, 1894), I, sec. II, 425; and George Thomas Brown, *The Pig: Its External and Internal Organisation* (London: George Philip, ca. 1900), 4–5. In an appendix to *A Natural History of Domesticated Mammals* (London: British Museum (Natural History), Cambridge: Cambridge University Press, 1987), Juliet Clutton-Brock accords no domestic species more than a single wild progenitor (196–197).

95. William Henry Flower, *The Horse: A Study in Natural History* (London: Kegan Paul, Trench, Trübner, 1891), 79.

96. Judith Neville Lytton, *Thoroughbred Racing Stock and Its Ancestors* (London: George Allen and Unwin, 1938), 39. For a discussion of the attractiveness of missing links to late-nineteenth-century readers, see Gillian Beer, *Forging the Missing Link: Interdisciplinary Stories* (Cambridge: Cambridge University Press, 1992).

97. Poster in Cambridgeshire Collection, Cambridge Central Library; Bartlett, "On Some Hybrid Bovine Animals," 399, 401.

98. Lyell, *Principles of Geology,* II, 49; Bigland, *Letters on Natural History,* 69; Peter Simon Pallas, *An Account of the Different Kinds of Sheep Found in the Russian Dominions and among the Tartar Hordes of Asia* (London: T. Chapman, 1794), 79.

99. Charles Hamilton Smith, *The Natural History of Horses. The Equidae or Genus Equus of Authors* (Edinburgh: W. H. Lizars, 1841), 344; Richardson, "Dogs," in Morton, *Cyclopedia of Agriculture,* I, 665.

100. Charles Kingsley, *The Water Babies* (1863; London: Penguin, 1995), 125.

101. Darwin, *Origin,* chap. 8, esp. 272–276.

102. W. C. Spooner, "On Cross Breeding" (London: W. Clowes, 1860), 21–22 (offprint from *Journal of the Royal Agricultural Society of England* 20, pt. 2).

103. Ridgeway, *Origin and Influence of the Thoroughbred Horse,* vii.

104. James Long, *The Book of the Pig: Its Selection, Breeding, Feeding, and Management* (London: L. Upcott Gill, 1886), 4; Vero Shaw, *The Illustrated Book of the Dog* (London: Cassell, Peter, Galpin, 1881), 53.

105. W. C. Spooner, "Breeding, Principles of," in Morton, *Cyclopedia of Agriculture,* I, 338.

106. Herman Biddell et al., *Heavy Horses: Breeds and Management* (London: Vinton, 1905), 15.

107. Thomas Bates and Thomas Bell, *The History of Improved Short-horn or Durham Cattle, and of the Kirklevington Herd* (Newcastle-upon-Tyne: Robert Redpath, 1871), 194.

108. Lewis Falley Allen, *History of the Short-horn Cattle: Their Origin, Progress and Present Condition* (Buffalo, N.Y.: The Author, 1874), 27.

109. "Cecil" [Cornelius Tongue], *Hints on Agriculture, Relative to Profitable Draining and Manuring; also the Comparative Merits of the Pure Breeds of Cattle and Sheep* (London: Thomas Cautley Newby, 1858), 118; William Housman, *Cattle: Breeds and Management* (London: Vinton, 1905), 127, 70, 47.

110. J. Rogers, *The Dog Fancier's Guide: Plain Instructions for Breeding and Managing the Several Varieties of Field, Sporting and Fancy Dogs* (London: Thomas Dean, n.d. [ca. 1850]), 26; W. F., "The Points of the Gordon Setter," *Kennel Gazette* 8 (1889), 39; "The Great Peterborough Hound Show," *Kennel Gazette* 2 (1881), 328; "Pillars of the Stud Book.—No. 11. The Irish Water Spaniel," *Kennel Gazette* 4 (1883), 469.

111. George Hanger, *Colonel George Hanger, to All Sportsmen, and Particularly to Farmers, and Game Keepers* (London: George Hanger, 1814), 47.

112. Gordon Stables, *The Practical Kennel Guide; with Plain Instructions How to Rear and Breed Dogs for Pleasure, Show, and Profit* (London: Cassell Petter and Galpin, 1877), 125.

113. Although it was clear to scientists, as it was to breeders, that sexual intercourse was necessary if the higher animals were to reproduce, there was no expert consensus until late in the nineteenth century about why this was so. On eighteenth- and nineteenth-century understandings of reproduction, see John Farley, *Gametes and Spores: Ideas About Sexual Reproduction 1750–1914* (Baltimore: Johns Hopkins University Press, 1982), chaps. 1 and 2, and Frederick B. Churchill, "Sex and the Single Organism: Biological Theories of Sexuality in Mid-Nineteenth Century," *Studies in the History of Biology* 3 (1979), 139–177.

114. M. Godine, "Comparative Influence of the Male and Female in Breeding," *Farrier and Naturalist* 1 (1828), 468.

115. "The Physiology of Breeding," *The Agricultural Magazine, Plough, and Farmers' Journal* (June 1855), 17.

116. E. Millais, "Influence; with special reference to that of the sire," in *Dog Owners' Annual* (1894), 153.

117. H. M. Gourrier, *The Law of Generation, Sexuality and Conception,* trans. and ed. Franklin Duane Pierce (Union Springs, N.Y.: Hygeia Publishing Company, 1886), 45; C. J. Davies, *The Kennel Handbook* (London: John Lane, 1905), 66.

118. Lord Morton to W. H. Wollaston, August 12, 1820, quoted in Ewart, *Penycuik Experiments,* 165–166.

119. G. H. Lewes, "Hereditary Influence, Animal and Human," *Westminster Review* 66 (1856), 85.

120. George M. Gould and Walter L. Pyle, *Anomalies and Curiosities of Medicine* (1896; New York: Bell, 1956), 87.

121. Alexander Harvey, *On a Remarkable Effect of Cross Breeding* (Edinburgh: William Blackwood, 1851), 4, 8.

122. Hugh Dalziel, *The St. Bernard; Its History, Points, Breeding, and Rearing* (London: L. Upcott Gill, 1889), 108.

123. Darwin, *Variation of Animals and Plants under Domestication,* I, 484; Reginald Orton, *On the Physiology of Breeding. Two Lectures Delivered to the Newcastle Farmer's Club* (Sunderland: The Times, 1855), 20. On Darwin's interest in telegony, see Mary M. Bartley, "Darwin and Domestication: Studies on Inheritance," *Journal of the History of Biology* 25 (1992), 327–329.

124. See Arthur Shipley, "Zebras, Horses, and Hybrids," *Quarterly Review* 190 (1899), 404–422, and Harvey, *On a Remarkable Effect of Cross Breeding,* for representative discussions of these alternative hypotheses.

125. Tegetmeier and Sutherland, *Horses,* 15.

126. The Earl Spencer, quoted in "Cecil" [Tongue], *Hints on Agriculture,* 171; Charles Darwin, *Darwin's Notebooks, 1836–1844: Geology, Transmutation of Species, Metaphysical Inquiries,* transcribed and edited by Paul H. Barrett, Peter J. Gautrey, Sandra Herbert, David Kohn, and Sydney Smith (Ithaca, N.Y.: Cornell University Press, 1981), 379.

127. Harvey, *On a Remarkable Effect of Cross Breeding,* 18.

128. Bates and Bell, *History of Improved Short-horn or Durham Cattle,* 86; Gould and Pyle, *Anomalies and Curiosities of Medicine,* 87.

129. Dalziel, *St. Bernard,* 109.

130. "Scotus," "Polled Angus or Aberdeenshire Cattle," in John Coleman, ed., *The Cattle of Great Britain: Being a Series of Articles on the Various Breeds of the United Kingdom, their Management, &c.* (London: "The Field," 1875), 103.

131. William Day, *The Horse: How to Breed and Rear Him* (London: Richard Bentley, 1890), 422.

132. C. J. Davies, "Fallacies of Breeding," *Dog Owners' Annual* (1901), 106.

133. For an overview of scientific ideas about telegony, see Richard W. Burkhardt, "Closing the Door on Lord Morton's Mare: The Rise and Fall of Telegony," *Studies in the History of Biology,* vol. 3, ed. William Coleman and Camille Limoges (Baltimore: Johns Hopkins University Press, 1979), 1–21.

134. Tegetmeier and Sutherland, *Horses,* 15.

135. Clutton-Brock, *Horse Power,* 47; Darwin, *Origin,* 165–167.

136. Shipley, "Zebras, Horses, and Hybrids," 410.

137. Ewart, *Penycuik Experiments*, 2, 3, 8.

138. See, for example, Ewart, *Penycuik Experiments*, lxix, lxvii, 63.

139. Davies, *Kennel Handbook*, 48.

140. Ewart, *Penycuik Experiments*, 161, 177.

141. Millais, "Influence," 153.

142. Everett Millais, *Two Problems of Reproduction* (Manchester: "Our Dogs," 1895), 18, 20.

143. Shaw, *Illustrated Book of the Dog*, 524.

144. William Taplin, *The Sportsman's Cabinet, or, A Correct Delineation of the Various Dogs Used in the Sports of the Field* (London, 1803), I, 27–28.

145. See, for example, Everard Home, *Lectures on Comparative Anatomy; in which are explained the Preparations in the Hunterian Collection*, III (London: Longman, Hurst, Rees, Orme, and Brown, 1823), 302; John Hunter, *Essays and Observations on Natural History, Anatomy, Physiology, Psychology, and Geology,* ed. Richard Owen (London: John Van Voorst, 1861), I, 195; and Gould and Pyle, *Anomalies and Curiosities of Medicine,* 46–48.

146. For a definitive account of this widely accepted conviction, see the first half of Marie-Helène Huet, *Monstrous Imagination* (Cambridge, Mass.: Harvard University Press, 1993). Dennis Todd has offered a more purely literary interpretation of similar material in *Imagining Monsters: Miscreations of the Self in Eighteenth-Century England* (Chicago: University of Chicago Press, 1995).

147. H. C. Brooke, "Some Foreign Dogs," *Dog Owners' Annual* (1897), 45.

148. William Youatt, *Cattle; Their Breeds, Management, and Diseases* (London: Baldwin and Craddock, 1834), 523.

149. Hugh Dalziel, *British Dogs: Their Varieties, History, Characteristics, Breeding, Management and Exhibition* (London: "Bazaar," 1879–1880), 461–463.

150. Shaw, *Illustrated Book of the Dog*, 275.

151. Vero Shaw, *British Horses Illustrated. With Brief Descriptive Notes on Every Native Breed* (London: Vinton, 1899), 5.

152. Shaw, *Illustrated Book of the Dog*, 274, 415.

153. James McDonald and James Sinclair, *History of Hereford Cattle* (London: Vinton, 1909), 31.

154. Shaw, *Illustrated Book of the Dog*, 155.

155. "Cattle," *Penny Magazine* 10 (1841), 282.

156. *The Flock Book of Suffolk Sheep*, I (Bury St. Edmunds: Suffolk Sheep Society, 1887), xiii.

157. *The Oxford Down Flock Book. Volume I. Rams Nos. 1 to 560* (London: Oxford Down Sheep Breeders' Association, 1889), 8.

158. Robert Wallace, *Farm Live Stock of Great Britain* (Edinburgh: Oliver and Boyd, 1885), 19.

159. On the development and acceptance of this understanding of pedigree, see Harriet Ritvo, "Possessing Mother Nature: Genetic Capital in Eighteenth-Century Britain,"

in John Brewer and Susan Staves, eds., *Early Modern Conceptions of Property* (New York: Routledge, 1995), 413–426.

160. Low, *Breeds of the Domestic Animals of the British Islands,* I, 9–10.

161. Youatt, *Cattle,* 524.

162. C. J. Davies, "Fallacies of Breeding," *Dog Owners' Annual* (1901), 109.

163. Biddell et al., *Heavy Horses,* 41.

164. Shipley, "Zebras, Horses, and Hybrids," 416.

165. Ewart, *Penycuik Experiments,* xli.

166. Shipley, "Zebras, Horses, and Hybrids," 416.

167. John Storer, *The Wild White Cattle of Great Britain: An Account of their Origin, History, and Present State,* ed. John Storer [son] (London: Cassell, Petter and Galpin, 1879), 215.

168. Millais, *Two Problems of Reproduction,* 8.

169. Lewes, "Hereditary Influence, Animal and Human," 86.

170. William Carr, *The History of the Rise and Progress of the Killerby, Studley and Warlaby Herds of Shorthorns* (London: William Ridgway, 1867), 21.

171. William Allison, *The British Thoroughbred Horse: His History and Breeding, Together with an Exposition of the Figure System* (London: Grant Richards, 1901), 16.

172. Professor Sheldon, "Shorthorns as Beef Producers," in John Watson, ed., *The Best Breeds of British Stock: A Practical Guide for Farmers and Owners of Live Stock in England and the Colonies* (London: W. Thacker, 1898), 1.

173. Jacob Wilson, "The Chillingham Wild Cattle," *The Land Magazine* (January 1899), 25–27.

174. Darwin, *Variation of Animals and Plants under Domestication,* II, 85; John Coleman, "Introductory," in Coleman, ed., *Cattle of Great Britain,* 1.

175. Cadwallader John Bates, *Thomas Bates and the Kirklevington Shorthorns: A Contribution to the History of Pure Durham Cattle* (Newcastle-upon-Tyne: Robert Redpath, 1897), 50.

176. George Culley, *Observations on Live Stock; Containing Hints for Choosing and Improving the Best Breeds of the Most Useful Kinds of Domestic Animals* (London, 1807), xi.

177. Vero Shaw, *How to Choose a Dog and How to Select a Puppy* (London: W. Thacker, 1897), 81.

178. Alfred Henry Huth, "Cross-Fertilisation of Plants, &c.," *Westminster Review* 108 (1877), 467, 483; George Henry Andrews, *Modern Husbandry: A Practical and Scientific Treatise on Agriculture* (London: Nathaniel Cooke, 1853), 162.

179. Huth, "Cross-Fertilisation of Plants," 467; G. W. Child, "Marriages of Consanguinity," *Westminster Review* 80 (1863), 89.

180. "Cecil" [Tongue], *Hints on Agriculture,* 122.

181. Everett Millais, "Basset Bloodhounds. Their Origin, Raison D'Etre and Value," *Dog Owners' Annual* (1897), 17.

182. Robert T. Vyner, *Notitia Venatica: A Treatise on Fox-Hunting. Embracing the General Management of Hounds and the Diseases of Dogs* (London, 1847), 24; George Tollet

to Charles Darwin, May 10, 1839, quoted in R. B. Freeman and P. J. Gautrey, "Darwin's *Questions about the Breeding of Animals,* with a Note on *Queries on Expression," Journal of the Society for the Bibliography of Natural History* 5 (1969), 322; W. D. Fox to Charles Darwin, ca. November 1838, in Darwin, *Correspondence,* II, 110; An Old Sportsman, "On Breeding Race Horses; With Remarks on Mr. Lawrence's History of the Horse," *Annals of Sporting and Fancy Gazette* 2 (1822), 2.

183. John Downing, *Private Catalogue of the Ashfield Herd of Pure-Bred Shorthorns* (London: Vinton, 1872), iii.

184. Day, *Horse,* 141, 137.

185. Long, *Book of the Pig,* 22, 24.

186. Blumenbach, *Anthropological Treatises,* 340.

187. See, for example, James Hannay, "Pedigree and Heraldry," *Westminster Review* 60 (1853), 46–47.

188. On the connection between human rank and animal pedigree, see Ritvo, *Animal Estate,* 52–63; Bates, *Thomas Bates and the Kirklevington Shorthorns,* 2.

189. *Davy's Devon Herd Book,* I (1851), iii.

190. Millais, *Two Problems of Reproduction,* 11.

191. Ritvo, *Animal Estate,* chap. 2.

192. Francis Galton, *Inquiries into Human Faculty and Its Development* (1883), quoted in Edward J. Larson, "The Rhetoric of Eugenics: Expert Authority and the Mental Deficiency Bill," *British Journal of the History of Science* 24 (1991), 45.

193. On the Enlightenment and post-Enlightenment history of human racial classification, see George W. Stocking, Jr., *Victorian Anthropology* (New York: Free Press, 1987); Michael Banton, *Racial Theories* (Cambridge: Cambridge University Press, 1987); Douglas A. Lorimer, *Color, Class and the Victorians* (Leicester: Leicester University Press, 1978), esp. chap. 7; Nancy Stepan, *The Idea of Race in Science: Great Britain, 1800–1960* (Hamden, Conn.: Archon Books, 1982); Martin Bernal, *Black Athena: The Afroasiatic Roots of Classical Civilization. Volume 1: The Fabrication of Ancient Greece, 1785–1985* (New Brunswick, N.J.: Rutgers University Press, 1987), chaps. 4 and 5; Anthony Pagden, *The Fall of Natural Man: The American Indian and the Origins of Comparative Ethnology* (Cambridge: Cambridge University Press, 1982); and Christine Bolt, *Victorian Attitudes to Race* (London: Routledge and Kegan Paul, 1971).

194. Charles Linné, *A General System of Nature,* ed. William Turton (London: Lackington, Allen, 1805), I, 9.

195. Johann Friedrich Blumenbach, *On the Natural Variety of Mankind,* III (1795), rpt. in Blumenbach, *Anthropological Treatises,* 209.

196. Buffon, *Barr's Buffon,* IV, 191, 324.

197. J. C. Prichard, "On the Relations of Ethnology to Other Branches of Knowledge," *Edinburgh New Philosophical Journal* 43 (1847), 311. According to Prichard, a facial angle of 90 degrees characterized "the ideal heads of Grecian gods."

198. Lavater, Sue and Co. (pseud.), *Lavater's Looking-Glass; or, Essays on the Face of Animated Nature, from Man to Plants* (London: Millar Ritchie, 1800), 20–21.

199. Thomas Henry Huxley, "On the Methods and Results of Ethnology," *Fortnightly Review* 1 (June 15, 1865), 257–262.

200. *Curiosities for the Ingenious. Selected from the Most Authentic Treasures of Nature, Science and Art* (London: Thomas Boys, 1822), 24–25.

201. Cynthia Eagle Russett, *Sexual Science: The Victorian Construction of Womanhood* (Cambridge, Mass.: Harvard University Press, 1989), 75; Francis Galton, *Hereditary Genius: An Inquiry into its Laws and Consequences* (London: Macmillan, 1892), 328.

202. "History of Man," *Westminster Review* 20 (1834), 188; for an extended discussion of the connection between such assessments and imperial relations, see Michael Adas, *Machines as the Measure of Men: Science, Technology and Ideologies of Western Dominance* (Ithaca, N.Y.: Cornell University Press, 1989).

203. John Lubbock, *On the Social and Religious Condition of the Lower Races of Man* (Liverpool: British Association, 1870), 37.

204. Thomas Laycock, *Naming and Classification of Mental Diseases and Defects* (London: J. E. Adland, 1863), 11 (reprinted from *Journal of Mental Science,* 1863.)

205. William Lawrence, *Lectures on Comparative Anatomy, Physiology, Zoology, and the Natural History of Man; delivered at the Royal College of Surgeons in the Years 1816, 1817, and 1818* (London: R. Carlile, 1823), 125.

206. White, *Account of the Regular Gradation in Man,* 134–135.

207. Linné, *System of Nature,* ed. Turton, 9; Blumenbach, *On the Natural Variety of Mankind,* in *Anthropological Treatises,* 264.

208. William Jardine, "Memoir of Camper," in William Jardine, *The Natural History of the Ruminating Animals, Part I, Containing Deer, Antelopes, Camels, &c* (Edinburgh: W. H. Lizars, 1835), 31.

209. John MacFadzean, "An Essay to Prove that there is but one Species of Man," Society for Investigating Natural History/Natural History Society: Papers Delivered, vol. 7 (1787–1788), 180, 182. Manuscript in Edinburgh University Library Special Collections.

210. James Cowles Prichard, *Researches into the Physical History of Man,* ed. George W. Stocking, Jr. (1813; Chicago: University of Chicago Press, 1973), 86.

211. Prichard, *Researches into the Physical History of Man,* iii.

212. Fleming, *Philosophy of Zoology,* II, 150.

213. William Henry Flower, "Presidential Address to the Department of Anthropology, 1881," in *Essays on Museums and Other Subjects Concerned with Natural History* (1889; New York: Books for Libraries Press, 1972), 240.

214. William Clift, "Skeletons and Preparations both Human and Zoological; forming part of the Anatomical Museum of Joshua Brookes Esq.—Theatre of Anatomy. Blenheim Street, Great Marlborough Street." Royal College of Surgeons ms. 176.h^A.21.

215. F. W. Newman, "*The Natural History of the Varieties of Man*. By R. G. Latham," *Prospective Review* 6 (1850), 449, 450.

216. William Henry Flower, "President's Address, Anthropological Institute of Great Britain and Ireland," *Journal of the Anthropological Institute of Great Britain and Ireland* 14 (1885), 391.

217. Stocking, *Victorian Anthropology*, 248–254; for the subsequent evolution of this debate, see Douglas Lorimer, "Theoretical Racism in Late-Victorian Anthropology, 1870–1900," *Victorian Studies* 31 (1988), 405–430.

218. Charles Darwin, *The Descent of Man and Selection in Relation to Sex*, ed. John Tyler Bonner and Robert M. May (1871; Princeton: Princeton University Press, 1981), I, 228, 235.

219. James Hunt, "On the Negro's Place in Nature," *Memoirs Read before the Anthropological Society of London* 1 (1863–64), 1, 51–52.

220. *Catalogue of a Great Variety of Natural and Artificial Curiosities, Now Exhibiting at the Large House, the Corner of Queen's Row, facing the Road, at Pimlico* (London, 1766), 4; G. M. Humphrey, *Analysis of the Physiological Series in the Gallery of the Museum of Comparative Anatomy* (Cambridge, 1866), 9.

221. Flower, "Presidential Address to the Department of Anthropology, 1881," 237.

222. Poster in John Johnson Collection, Bodleian Library, Oxford. They were actually fakes, subsequently exposed as "a travelling tinker, and his *pal*, a gipsey." "Nottinghamshire. *Natural History*," *Annals of Sporting and Fancy Gazette* 2 (1822), 408.

223. *Nature* (May 12, 1882), cited in Martin Howard, *Victorian Grotesque: An Illustrated Excursion into Medical Curiosities, Freaks and Abnormalities—Principally of the Victorian Age* (London: Jupiter Books, 1977), 56–57.

224. Paul Greenhalgh, *Ephemeral Vistas: The Expositions Universelles, Great Exhibitions and World's Fairs, 1851–1939* (Manchester: Manchester University Press, 1988), 85.

225. Benedict Burton, "The Anthropology of World's Fairs," in Benedict Burton et al., *The Anthropology of World's Fairs: San Francisco's Panama Pacific International Exposition of 1915* (London: Scolar Press, 1983), 46.

226. Knox, *Races of Men*, 44–45.

227. William Clark, *Catalogue of the Osteological Portions of Specimens Contained in the Anatomical Museum of the University of Cambridge* (Cambridge: Cambridge University Press, 1862), 109–116.

228. Robert Knox, *Great Artists and Great Anatomists; A Biographical and Philosophical Study* (London: John Van Voorst, 1852), 18–19.

229. Knox, *Races of Men*, 50, 57, 10, 54.

230. John Beddoe, *The Races of Britain: A Contribution to the Anthropology of Western Europe* (Bristol: J. W. Arrowsmith, 1885), 9, 11.

231. Galton, *Hereditary Genius*, 328.

232. For an extended discussion of the metaphorical implications of human hybridization in the context of imperialism, see Robert J. C. Young, *Colonial Desire: Hybridity in Theory, Culture and Race* (London: Routledge, 1995).

233. Knox, *Races of Men*, 499, 506.

234. S. Anderson Smith, "The Degeneration of Race," *Lancet* 1 (February 23, 1861), 202; "The War of Races," *New Monthly Magazine* 90 (1850), 80.

235. See, for example, White, *Account of the Regular Gradation in Man,* 129; Hunt, "On the Negro's Place in Nature," 25; and Broca, *On the Phenomena of Hybridity in the Genus Homo,* 36–37.

236. Broca, *On the Phenomena of Hybridity in the Genus Homo,* 47; Stocking, *Victorian Anthropology,* 283.

237. Darwin, *Descent of Man,* I, 225; Knox, *Races of Men,* 67. For representative analyses of the mulattos and mestizos of the Americas, see Blumenbach, *On the Natural Variety of Mankind,* 216–218, and W. B. Stevenson, *Narrative of Twenty Years Residence in South America* (1825), reproduced in Mary Louise Pratt, *Imperial Eyes: Travel Writing and Transculturation* (London: Routledge, 1992), 152.

238. Darwin, *Variation of Animals and Plants under Domestication,* II, 63.

239. "Types of Mankind," *Westminster Review* 65 (1856), 380; *The Natural History of Man* (London: William Darton, 1835), 160.

240. *Dr. Darwin. New Edition* (London: W. S. Fortey, n.d.).

241. *The Complete Farmer: Or a General Dictionary of Husbandry in All Its Branches by a Society of Gentlemen* (London: S. Crowder, 1766), not paginated; James Watson Roberts, "Of the Degeneration of Animals," Society for Investigating Natural History/Natural History Society: Papers Delivered, vol. IV (1785–1785), 108, manuscript in Edinburgh University Library, Special Collections; Lyell, *Principles of Geology,* II, 64.

4. Out of Bounds

1. G. H. Lewes, "Hereditary Influence, Animal and Human," *Westminster Review* 66 (1856), 76.

2. G. W. Child, "Marriages of Consanguinity," *Westminster Review,* new ser., 24 (1863), 99.

3. William Henry Flower, *Fashion in Deformity as Illustrated in the Customs of Barbarians and Civilised Races* (London: Macmillan, 1881), 6, 1, 4–5, 63, 78–80.

4. Charles Dickens, *Our Mutual Friend* (1865; rpt., New York: Modern Library, 1960), 85–86.

5. See the discussion of the platypus in Chapter 1, above. William B. Ashworth, Jr., "Remarkable Humans and Singular Beasts," in Joy Kenseth, ed., *The Age of the Marvelous* (Hanover, N.H.: Hood Museum of Art, Dartmouth College, 1991), 116–117.

6. John Johnson Collection, Bodleian Library, Oxford.

7. Poster in the John Johnson Collection, Animals on Show 2.

8. "Cumberland. *Monstrous animal production,*" *Annals of Sporting and Fancy Gazette* 1 (1822), 413–414.

9. Thomas Pennant, *The British Zoology* (London: J. and J. March, 1766), 6.

10. David Low, *The Breeds of the Domestic Animals of the British Islands* (London: Longman, Orme, Brown, Green and Longmans, 1842), iv.

11. "Professor Monro's [Secundus] Lectures, Surgery, Comparative Anatomy, Optics," Notes of J. C. Hope, 1786, vol. I, not paginated, Edinburgh University Special Collections; Busick Harwood, *A Synopsis of a Course of Lectures on the Philosophy of Natural History and the Comparative Structure of Plants and Animals* (Cambridge: Francis Hodson, 1812), 9.

12. James Wilson, "Essay on the Origin and Natural History of Domestic Mammals," *Farrier and Naturalist* 2 (1829), 292.

13. Robert Knox, *The Races of Men: A Philosophical Enquiry into the Influence of Race over the Destinies of Nations* (London: Henry Renshaw, 1862), 88.

14. Flower, *Fashion in Deformity*, 85.

15. Thomas Boulton, "On a Case of Monstrosity," *Lancet*, May 7, 1864, 517.

16. Alexander Walker, *Intermarriage: Or the mode in which, and the causes why, Beauty, Health and Intellect, Result from Certain Unions, and Deformity, Disease and Insanity, from Others* (New York: J. and H. G. Langley, 1839), 131–137.

17. *Notes and Queries on Anthropology, for the Use of Travellers and Residents in Uncivilized Lands* (London: Edward Stanford, 1874), 15.

18. For a comprehensive survey of London exhibits, see Richard D. Altick, *The Shows of London: A Panoramic History of Exhibitions, 1600–1862* (Cambridge, Mass.: Harvard University Press, 1978), esp. chaps. 3 and 19. Other accounts of monstrosities on display in the eighteenth and nineteenth centuries include C. J. S. Thompson, *The Mystery and Lore of Monsters, with Accounts of Some Giants, Dwarfs and Prodigies* (London: Williams and Norgate, 1930); David Murray, *Museums, their History and their Use*, I (Glasgow: James MacLehose, 1904); Robert Bogdan, *Freak Show: Presenting Human Oddities for Amusement and Profit* (Chicago: University of Chicago Press, 1988); and Martin Howard, *Victorian Grotesque: An Illustrated Excursion into Medical Curiosities, Freaks and Abnormalities—Principally of the Victorian Age* (London: Jupiter Books, 1977).

19. "Living Skeleton," *Lancet* 8 (1825), 178–179.

20. Kenneth Grahame, "Sanger and his Times," introduction to "Lord" George Sanger, *Seventy Years A Showman* (1910; rpt., London: J. M. Dent, 1926), 17–18.

21. Notices and advertisements quoted in Edward J. Wood, *Giants and Dwarfs* (1868; rpt., London: Folcroft Library, 1974), 129, 150, 309–310.

22. R. T. Gunther, *Early Science in Cambridge* (Oxford: Oxford University Press, 1937), 342; Thomas Frost, *The Old Showmen and the Old London Fairs* (London: Chatto and Windus, 1881), 204.

23. Howard, *Victorian Grotesque*, 48.

24. *A Catalogue of the Cabinet of Birds, and Other Curiosities, Now Exhibiting at the New House* (London: n.p., 1769), 12, introduction (not paginated).

25. A. Ella, *Visits to the Leverian Museum; Containing an Account of Several of its Principle Curiosities, Both of Nature and Art: Intended for the Instruction of Young Persons in the First Principles of Natural History* (London: Tabart, 1805), iv–v.

26. *A Descriptive Catalogue of Rackstrow's Museum* (London: n.p., 1782), 18–19, 20.

27. A. E. Gunther, *The Founders of Science at the British Museum, 1753–1900* (Halesworth, Suffolk: Halesworth Press, 1980), 23; "Drawings of Animals, Monsters, Skeletons, etc.," British Library Manuscripts, Sloane 5220.

28. R. Lydekker, "Domesticated Animals, Hybrids, and Abnormalities," in *The History of the Collections Contained in the Natural History Departments of the British Museum*, vol. II, *Separate Accounts of the Several Collections Included in the Department of Zoology* (London: British Museum, 1906), 73; [R. Lydekker], *Guide to the Specimens of the Horse Family (Equidae Exhibited in the Department of Zoology, British Museum (Natural History)* (London: British Museum, 1907), 9–10.

29. Frank Woolnough, *A Guide to the Ipswich Museum* (Ipswich: F. Pawsey, 1895), 16–17.

30. Dudley Wilson, *Signs and Portents: Monstrous Births from the Middle Ages to the Enlightenment* (Routledge: London, 1993), 114.

31. Gunther, *Early Science in Cambridge*, 307.

32. M. F. Ashley Montagu, *Edward Tyson, M.D., F.R.S., 1650–1708, and the Rise of Human and Comparative Anatomy in England: A Study in the History of Science* (Philadelphia: American Philosophical Society, 1943), 344; *Royal Society: Early Letters and Classified Papers, 1660–1740* (microfilm) (Bethesda, Md.: University Publications of America, 1990), Reel 18, "Monsters; Longevity."

33. "Skeletons and Preparations both Human and Zoological; forming part of the Anatomical Museum of Joshua Brookes Esq.," manuscripts in Royal College of Surgeons Library, London, 276.h^A.21.

34. William Clift, "Articles Purchased at Mr. Brookes's Sale, July 1828," Royal College of Surgeons Library, manuscript 275.h.15(5); William Henry Flower, *Essays on Museums and Other Subjects Connected with Natural History* (1898; rpt., Freeport, N.Y.: Books for Libraries Press, 1972), 90.

35. Cambridge University Register 39.13, "Professor of Anatomy.," Items 43 and 58, Cambridge University Archives; Cambridge University Museum of Zoology, "Additions to the Museum," vol. I (1867–1902), 155–156, CUMZ Archives.

36. Richard Owen, *The Hunterian Lectures in Comparative Anatomy, May and June 1837*, ed. Phillip Reid Sloan (Chicago: University of Chicago Press, 1992), 185; Hugh Miller, *The Foot-prints of the Creator; or, the Asterolepis of Stromness*, quoted in James G. Paradis, "The Natural Historian as Antiquary of the World: Hugh Miller and the Hieroglyphs of Geology," in Michael Shortland, ed., *Hugh Miller and the Controversies of Victorian Science* (Oxford: Oxford University Press, 1996), 140.

37. Frederick Edward Hulme, *Natural History Lore and Legend* (London: Quaritch, 1895), 62.

38. George M. Gould and Walter L. Pyle, *Anomalies and Curiosities of Medicine* (1896; New York: Bell, 1956), 1.

39. For discussions of this literature, see Marie-Hélène Huet, *Monstrous Imagination* (Cambridge, Mass.: Harvard University Press, 1993), chap. 1, and Katherine Park and Lorraine J. Daston, "Unnatural Conceptions: The Study of Monsters in

Sixteenth- and Seventeenth-Century France and England," *Past and Present* 92 (1981), 25–35.

40. Ulisse Aldrovandi, *Monstrorum Historia* (Bologna: Nicolai Tebaldini, 1642), 10, 17, 21, 25, 353, 366, 401, 411, 416–422.

41. Ambroise Paré, *On Monsters and Marvels*, trans. and ed. Janis L. Pallister (1573; Chicago: University of Chicago Press, 1982), 3. The first English edition appeared in 1634.

42. Park and Daston, "Unnatural Conceptions," 54.

43. John Kidd, *An Introductory Lecture to a Course in Comparative Anatomy, Illustrative of Paley's Natural Theology* (Oxford: John Kidd, 1826), 36.

44. "Hen with a Human Face," *Lancet* (1829–1830), 105–106.

45. John North, "A Lecture on Monstrosities," *Lancet* (March 7, 1840), 857.

46. Hulme, *Natural History Lore and Legend*, 61–62.

47. *Oxford English Dictionary;* Huet, *Monstrous Imagination*, 108. For a brief account of Etienne Geoffroy Saint-Hilaire's views on the laws governing the development of monsters and his son Isidore Geoffroy Saint-Hilaire's attempts to classify them, see (in addition to Huet) Jean-Louis Fischer, *Monstres: Histoire du Corps et de ses Défauts* (Paris: Syros-Alternatives, 1991), 88–96.

48. Thompson, *Mystery and Lore of Monsters*, 70.

49. John Hunter, *Essays and Observations on Natural History, Anatomy, Physiology, Psychology, and Geology*, ed. Richard Owen (London: John Van Voorst, 1861), I, 248, 240; Owen, *Hunterian Lectures*, 185.

50. Quoted in Owen E. Clark, "The Contributions of J. F. Meckel, the Younger, to the Science of Teratology," *Journal of the History of Medicine and Allied Sciences* 24 (1969), 317.

51. Huet, *Monstrous Imagination*, 108 ff.

52. Georges Louis Leclerc, Comte de Buffon, *Barr's Buffon. Buffon's Natural History* (London: H. D. Symonds, 1797), VIII, 29; W. Webb, "On the Effects of Domestication on Animals," Society for Investigating Natural History/Natural History Society, Papers Delivered (Minutes in the Edinburgh University Library Special Collections), XII (1793–1794), 397.

53. John Fleming, *The Philosophy of Zoology; or A General View of the Structure, Functions, and Classification of Animals* (Edinburgh: Archibald Constable, 1822), II, 150.

54. Charles Darwin, *Charles Darwin's Notebooks, 1836–1844: Geology, Transmutation of Species, Metaphysical Inquiries*, transcribed and ed. Paul H. Barrett et al. (Ithaca, N.Y.: Cornell University Press, 1987), 311.

55. Holmes Coote, "On the Nature and Treatment of Deformities," *Lancet* (February 25, 1860), 191.

56. Robert Fox, "Notices of Monstrosities," *Lancet* (1840), 471.

57. W. Vrolik, "Teratology," in Robert Bentley Todd, ed., *The Cyclopedia of Anatomy and Physiology* (London: Sherwood, Gilbert, and Piper, 1836–1859), IV, 946.

58. John Cleland, "Teratology, Speculative and Causal, and the Classification of

Anomalies," in John Cleland et al., *Memoirs and Memoranda in Anatomy* (London: Williams and Norgate, 1889), I, 127.

59. William Lawrence, *Lectures on Comparative Anatomy, Physiology, Zoology, and the Natural History of Man; delivered at the Royal College of Surgeons in the Years 1816, 1817, and 1818* (London: R. Carlile, 1823), 421.

60. North, "Lecture on Monstrosities," 858.

61. North, "Lecture on Monstrosities," 858–859.

62. Clark, "Contributions of J. F. Meckel," 316–317; E. S. Russell, *Form and Function: A Contribution to the History of Animal Morphology* (1916; Chicago: University of Chicago Press, 1982), 93–94.

63. Darwin, *Charles Darwin's Notebooks*, 199, 259.

64. Edward Blyth to Charles Darwin, September 10 or October 7, 1855, *The Correspondence of Charles Darwin. Volume 5. 1851–1855,* ed. Frederick Burkhardt et al. (Cambridge: Cambridge University Press, 1989), 445.

65. Charles Darwin, *On the Origin of Species*, ed. Ernst Mayr (1859; Cambridge, Mass.: Harvard University Press, 1964), chap. 1.

66. See Martin J. S. Rudwick, *Scenes from Deep Time: Early Pictorial Representations of the Prehistoric World* (Chicago: University of Chicago Press, 1992), and A. Bowdoin Van Riper, *Men among the Mammoths: Victorian Science and the Discovery of Human Prehistory* (Chicago: University of Chicago Press, 1993), for recent accounts of this emergent consensus.

67. Evelleen Richards, "A Political Anatomy of Monsters, Hopeful and Otherwise: Teratogeny, Transcendentalism, and Evolutionary Theorizing," *Isis* 85 (1994), 377–411, esp. 399. Her "A Question of Property Rights: Richard Owen's Evolutionism Reassessed," *British Journal of the History of Science* 20 (1987), 129–171, also deals powerfully with Owen's use of teratology in elaborating his ideas about evolution.

68. Darwin, *Origin*, 424.

69. See, for example, Charles Darwin, *The Variation of Animals and Plants under Domestication* (New York: Orange Judd, 1868), II, 76–79, and Charles Darwin, *The Descent of Man, and Selection in Relation to Sex* (1871; rpt., Princeton, N.J.: Princeton University Press, 1981), I, 223–224.

70. Chris Baldick, *In Frankenstein's Shadow: Myth, Monstrosity, and Nineteenth Century Writing* (Oxford: Oxford University Press, 1987), 11–12.

71. *England's New Wonders; Or, Four Strange and Amazing Relations That Have Lately Come to Pass in England* (London: J. Blore, 1697), 4–10.

72. *The Works of Aristotle, the Famous Philosopher . . .* (London: "Printed for the Booksellers," 1802), I, 50–51.

73. Charles Lyell, *Principles of Geology* (1832; rpt., Chicago: University of Chicago Press, 1991), II, 63–64.

74. Thomas Litchfield, "An Account of a Monstrosity after a Painful Labour," *Lancet* (1850), 50.

75. See the entries for *monster* and *monstrous* in the *Oxford English Dictionary*.

76. Poster in John Johnson Collection, Human Freaks 2.

77. *Biographical Sketch of Millie Christine, the Two-Headed Nightingale* (n.p., n.d.), 19. John Johnson Collection, Human Freaks 2.

78. Fliers in the John Johnson Collection, Human Freaks 2.

79. Izett W. Anderson, "'The Turtle Woman' of Demerara," *Lancet* (November 9, 1867), 578.

80. Fox, "Notices of Monstrosities," 471.

81. Gould and Pyle, *Anomalies and Curiosities of Medicine*, 2.

82. For a detailed, if undoubtedly partial, listing of the stream of giants on display in eighteenth- and nineteenth-century Britain, see Wood, *Giants and Dwarfs*, 125–231.

83. "Fairburn's ACCURATE PORTRAITS of the TWO most CORPULENT ENGLISH-MEN ever known, with a COMPARATIVE ACCOUNT of their EXTRAORDI-NARY PERSONS and MANNERS," (London: John Fairburn, 1806), broadside in John Johnson Collection; poster in John Johnson Collection.

84. James Y. Simpson, *The Siamese Twins, Chang and Eng . . .* (London: J. W. Last, ca. 1870), 29; Richard Aldington, *The Strange Life of Charles Waterton, 1782–1865* (New York: Duell, Sloan and Pearce, 1949), 200.

85. "The Irish Giant," Information Sheet No. 3, Hunterian Museum, Royal College of Surgeons of England, n.d.

86. Gould and Pyle, *Anomalies and Curiosities of Medicine*, 331.

87. For accounts of Byrne's career, see "The Irish Giant," Hunterian Museum; Wood, *Giants and Dwarfs*, 158–164; Gould and Pyle, *Anomalies and Curiosities of Medicine*, 330–331; and Howard, *Victorian Grotesque*, 97–98.

88. This account relies on the definitive research of Jan Bondeson in "Caroline Crachami, the Sicilian Fairy: A Case of Bird-Headed Dwarfism," *American Journal of Medical Genetics* 44 (1992), 210–219, and "Caroline Crachami, the Sicilian Fairy: A Further Note," *American Journal of Medical Genetics* 46 (1993), 471; and on Altick's detailed relation of the unsavory circumstances surrounding the disposal of Crachami's body in *Shows of London*, 257–260.

89. The modern distinction between perfectly formed but diminutive midgets and disproportioned dwarfs was not current in the eighteenth and nineteenth centuries; according to the *Oxford English Dictionary*, the term *midget* did not appear in print until the second half of the nineteenth century. For another kind of distinction, between dwarf shows in fairs and dwarf shows in circuses, see Yoram S. Carmeli, "From Curiosity to Prop—A Note on the Changing Cultural Significances of Dwarves' Presentations in Britain," *Journal of Popular Culture* 26 (1992), 69–80.

90. Poster, John Johnson Collection, Human Freaks 1.

91. Wood, *Giants and Dwarfs*, 394.

92. Posters in the John Johnson Collection, Human Freaks 1.

93. Henry Davies, "An Account of Two Labours of a Dwarf," *Lancet* 2 (1847), 488–491.

94. Lavater, Sue, and Co. [pseud.], *Lavater's Looking Glass; or, Essays on the Face of Animated Nature, from Man to Plants* (London: Millar Ritchie, 1800), 63.

95. See, for example, Gould and Pyle, *Anomalies and Curiosities of Medicine,* 233–235.

96. *An Historical Miscellany of the Curiosities and Rarities in Nature and Art* (London: Champante and Whitrow, ca. 1800), III, 289.

97. Circular, ca. 1815, in John Johnson Collection, Human Freaks 1.

98. Poster in Countway Library Archives, f.QL.991.T55, Harvard Medical School, Boston.

99. Poster in John Johnson Collection, Animals on Show 2.

100. Wernerian Natural History Society Minute Books, I, 188, Edinburgh University Library Special Collections.

101. "An Extraordinary Freak of Nature," broadside in the John Johnson Collection, Human Freaks 1.

102. "Anatomy of a Monster," *Lancet* (1843), 586.

103. "Acephalous Monster at the Full Period of Pregnancy," *Lancet* (December 1848), 696.

104. For example, "Acephalous Monster," *Lancet* (August 3, 1861), 125, described a birth at the Paris Maternity Hospital.

105. "Anatomy of a Monster," *Lancet* (1843), 586.

106. For example, Robert Allan, "Dissection of a Human Astomatous Cyclops," *Lancet* (1848), 221.

107. George E. Jeaffreson, "An Acephalous Monster," *Lancet* 17 (1866), 568.

108. Allan, "Dissection of a Human Astomatous Cyclops," 222.

109. J. D. Hulme, "Another Monster," *Lancet* (October 22, 1864), 481.

110. John Scott, "A Monoculous Male Foetus," *Lancet* (June 14, 1862), 633.

111. Jeaffreson, "Acephalous Monster," 568.

112. W. Andrew, "Monstrous Female Foetus," *Lancet* (1839), 333.

113. "An Extraordinary Freak of Nature," broadside in John Johnson Collection, Human Freaks 1; Scott, "Monoculous Male Foetus," 633.

114. For example, a double pig engraved by the young Albrecht Dürer in 1496 was denounced as a portent of the coming of Antichrist. Kenseth, ed., *Age of the Marvelous,* 325.

115. "Lusus naturae," *Annals of Sporting and Fancy Gazette,* 1 (1822), 419.

116. "Another cow," *Annals of Sporting and Fancy Gazette* 9 (1826), 314.

117. *The Fancy (or True Sportsman's) Guide* 2 (1822–26), 45.

118. Posters in the John Johnson Collection, Animals on Show 2.

119. See, for example, "Hereditary Formations," *Lancet* (September 5, 1857), 258.

120. Simpson, *Siamese Twins,* 1.

121. Gould and Pyle, *Anomalies and Curiosities of Medicine,* 168.

122. Simpson, *Siamese Twins.*

123. See, for example, "A Monster Birth," *Lancet* (1846), 667; and "Double Monstrosity," *Lancet* (November 11, 1865), 548.

124. Fox, "Notices of Monstrosities," 471.

125. Jan Bondeson, "The Isle-Brewers Conjoined Twins of 1680," *Journal of the Royal Society of Medicine* 86 (1993), 106–107.

126. "Human Monster with Two Heads," *Lancet* (1829–1830), 194–195.

127. Gould and Pyle, *Anomalies and Curiosities of Medicine*, 184–185.

128. For a survey of celebrated double monsters, see Gould and Pyle, *Anomalies and Curiosities of Medicine*, chap. 5.

129. Simpson, *Siamese Twins*, 20–21.

130. *Biographical Sketch of Millie Christine*, 12.

131. Simpson, *Siamese Twins*, 27, 12–13.

132. *Biographical Sketch of Millie Christine*, 6.

133. Cleland, "Teratology, Speculative and Causal," 131.

134. James Y. Simpson, "Hermaphroditism," in Todd, ed., *Cyclopedia of Anatomy and Physiology*, II, 684.

135. Robert Knox, *Contributions to Anatomy and Physiology* (London: Wilson and Ogilvy, 1843), 35.

136. John Hill, *A Review of the Works of the Royal Society of London; Containing Animadversions on such of the Papers as deserve Particular Observation* (London: R. Griffiths, 1751), 97.

137. D. Allen, "Of a Hermaphrodite" (January 16, 1667/8), *Royal Society: Early Letters and Classified Papers, 1660–1740*, Reel 19: "Monsters; Longevity," July 1, 1663–November 8, 1772; Wernerian Natural History Society Minute Books, 1808–1858, Edinburgh University Library Special Collections DC 2.55–56, II, 337.

138. Thomas Coltman to Joseph Banks, October 27, 1787, and Thomas Andrew Knight to Joseph Banks, December 17, 1801, in Harold B. Carter, ed., *The Sheep and Wool Correspondence of Sir Joseph Banks, 1781–1820* (New South Wales: Library Council of New South Wales, 1979), 130, 357.

139. Everard Home to K. Balfour, October 6, 1802, in Royal College of Surgeons Archive, 275.h.4; *Catalogue of the Contents of the Museum of the Royal College of Surgeons in London. Part v. Comprehending the Preparations of Monsters and Malformed Parts, in Spirit, and in a Dried State* (London: Richard Taylor, 1831), 60 ff.

140. "Daily Register: Objects illustrative of Medical Science added to the Anatomical Museum of the University of Glasgow from October 1848," 31 (University of Glasgow Special Collections, MR 49/4).

141. "Medical Society of London," *Lancet* (1838), 915.

142. "Charing-Cross Hospital. Remarkable Arrest of Development in the Organs of Generation and in the Urinary Apparatus," *Lancet* (1840), 937–938; "Charing Cross Hospital. Dissection of the Parts of Generation of a Supposed Woman. Clinical Remarks," *Lancet* (1842), 374; Edward Smith, "Singular Case of Malformation of the Sexual Organs, with Absence of the Normal Urethra," *Lancet* (1844), 183.

143. "Description of an Hermaphrodite Orang-Outang, Lately Living in Philadelphia," *Lancet* (1835–36), 963–965.

144. Reproduced in Wilson, *Signs and Portents*, 93.

145. "For the Satisfaction of the CURIOUS . . . ," broadside in Countway Library, RC.883.F74.

146. "A Human Tripod," *Lancet* (September 19, 1868), 397.

147. Fleming, *Philosophy of Zoology*, II, 150.

148. Thomas Pennant, *History of Quadrupeds* (London: B. and J. White, 1793), I, 14.

149. Thomas Hawkins, *The Book of the Great Sea Dragons, Ichthyosauri and Plesiosauri, . . . Gedolim Taninim of Moses. Extinct Monsters of the Ancient Earth* (London: William Pickering, 1840), 14.

150. A. T. Thomson, "Lectures on Medical Jurisprudence . . . Lecture IX," *Lancet* (1836), 351.

151. Knox, *Contributions to Anatomy and Physiology*, 42–43.

152. *Hermaphrodite* (London: A. Steward, 1750), broadside in Pathological Preparations Box, Royal College of Surgeons.

153. "Apparent Hermaphroditism. Sexual Connexion through the Canal of the Urethra," *Lancet* (1837), 15.

154. "Westminster Medical Society. Hermaphrodism," *Lancet* (1829), 181–182.

155. William Loney, "On a Case of Hermaphroditism," *Lancet* (1856), 624–625.

156. G. F. Girdwood, "On Hermaphroditism," *Lancet* (1859), 639–640. The unhappy life of a similarly endowed French hermaphrodite at about the same period has been documented in Michel Foucault, ed., *Herculine Barbin dite Alexina B.* (Paris: Gallimard, 1978). For an account of the medical inclination, which strengthened over the course of the nineteenth century, to consign apparently ambiguous individuals to one of the two conventional sexes, see Alice Domurat Dreger, "Doubtful Sex: The Fate of the Hermaphrodite in Victorian Medicine," *Victorian Studies* 38 (1995), 335–370.

157. Girdwood, "On Hermaphroditism," 640–641.

158. *Hermaphrodite* broadside, not paginated.

159. For accounts of well-known eighteenth- and nineteenth-century hermaphrodites, see Gould and Pyle, *Anomalies and Curiosities of Medicine*, 206–212; and Lynne Friedli, "'Passing Women'—A Study of Gender Boundaries in the Eighteenth Century," in G. S. Rousseau and Roy Porter, eds., *Sexual Underworlds of the Enlightenment* (Manchester: University of Manchester Press, 1987), 246–250.

160. Gilbert Murray, "The Ayrshire Breed of Cattle," in John Coleman, ed., *The Cattle of Great Britain: Being a Series of Articles on the Various Breeds of Cattle of the United Kingdom, their Management, &c.* (London: "The Field," 1875), 108.

161. Terry Castle, *Masquerade and Civilization: The Carnivalesque in Eighteenth-Century English Culture and Fiction* (Palo Alto: Stanford University Press, 1986), 36, 46; Fleming, *Philosophy of Zoology*, II, 149–150; R. Aldington, *The Strange Life of Charles Waterton, 1782–1865* (London: Evans, 1949), 185.

162. "Minute Book of the Zoological Club of the Linnean Society," February 27, 1827, Manuscript Drawer 36, Linnean Society Library, London; Walker, *Intermarriage*, 46, 44.

163. It was subsequently established that this phenomenon occurs because the embryonic male hormones of the twin bull calf inhibit the development of the female reproductive organs. W. E. Petersen, *Dairy Science*, ed. R. W. Gregory (Chicago: J. B. Lippincott, 1939), 221.

164. John Hunter, *Observations on Certain Parts of the Animal OEconomy* (London, 1786), 49.

165. Harwood, *Synopsis,* 13; Simpson, "Hermaphroditism," 719; George Vasey, *A Monograph of the Genus Bos. The Natural History of Bulls, Bisons, and Buffaloes* (London: John Russell Smith, 1857), 156.

166. William M'Combie, *Cattle and Cattle-Breeders* (Edinburgh: William Blackwood, 1867), 178–179.

167. Letter from James Grierson, *Farmer's Journal* 5 (1812), 193.

168. Nathaniel Highmore, *Case of a Foetus found in the Abdomen of a Young Man, at Sherborne, in Dorsetshire* (London: Longman, Hurst, Rees, Orme and Brown, 1815), 13, 19, 30.

169. "Skeletons and Preparations both Human and Zoological; forming part of the Anatomical Museum of Joshua Brookes, Esq.," Royal College of Surgeons Manuscripts 276.h^A.21 (in handwriting of William Clift); Dr. Blundell, "Lectures on the Gravid Uterus, and on the Diseases of Women and Children . . . Lecture VI," *Lancet* (1829), 261.

170. Simpson, "Hermaphroditism," 736.

171. Poster in John Johnson Collection; *An Interesting Treatise on The Marvellous Indian Boy Lalloo Brought to this Country by M. D. Francis* (Leicester: W. Willson, n.d.).

172. John Ashton, *Curious Creatures in Zoology* (New York: Cassell, 1890), 47; Gould and Pyle also list a number of Victorian examples in *Anomalies and Curiosities of Medicine,* 228–232.

173. Poster in John Johnson Collection; Jan Bondeson and A. E. W. Miles, "Julia Pastrana, the Nondescript: An Example of Congenital, Generalized Hypertrichosis Terminalis with Gingival Hyperplasia," *American Journal of Medical Genetics* 74 (1993), 199.

174. J. Z. Laurence, "A Short Account of the Bearded and Hairy Female now being exhibited at the Regent-Gallery," *Lancet* (1859), 48.

175. Bondeson and Miles, "Julia Pastrana, the Nondescript," 203–204.

176. T., "The Hare," *Annals of Sporting and Fancy Gazette* 4 (1823), 237–238.

177. Thomas Bewick, *A General History of Quadrupeds* (Newcastle upon Tyne: T. Bewick, 1824), 271.

178. Don E. Wilson and DeeAnn M. Reeder, *Mammal Species of the World: A Taxonomic and Geographic Reference,* 2d ed. (Washington, D.C.: Smithsonian Institution Press, 1993), 344.

179. Hunter, *Observations on Certain Parts of the Animal OEconomy,* 46; Hunter, *Essays and Observations on Natural History,* I, 184.

180. Thomson, "Lectures on Medical Jurisprudence," 350–351; "Description of an Hermaphrodite Orang-outang," 964.

181. H. W. Dewhurst, "Remarks on Hermaphrodism, in Man and Animals," *Hippiatrist (Farrier and Naturalist)* 3 (1830), 262.

182. H. M. Gourrier, *The Laws of Generation, Sexuality and Conception,* trans. and ed.

Franklin Duane Pierce (Union Springs, N.Y.: Hygeia Publishing Company, 1886), 28.

183. Knox, *Contributions to Anatomy and Physiology,* 35.

184. Girdwood, "On Hermaphroditism," 641. On the eighteenth- and nineteenth-century debates about the possibility of human hermaphrodism, see Friedli, "'Passing Women'," 246–257.

185. Mark Jones, ed., with Paul Craddock and Nicolas Barker, *Fake? The Art of Deception* (Berkeley, Calif.: University of California Press, 1990), 87; Pennant, *History of Quadrupeds,* II, 99; Daniel S. Simberloff, "A Funny Thing Happened on the Way to the Taxidermist," *Natural History* (1987), 50–55. Simberloff suggests that a papilloma virus can produce hornlike excrescences in lagomorphs that might have given rise to the notion of the jackalope; no such naturalistic explanation has been offered for the furry fish.

186. Hill, *Review of the Works of the Royal Society,* 143; H. M. Sinclair and A. H. T. Robb-Smith, *A Short History of Anatomical Teaching in Oxford* (Oxford: Oxford University Press, 1950), 23.

187. George Henry Millar, *A New, Complete, and Universal Body, or System of Natural History . . .* (London: Alexander Hogg, 1785), 18.

188. Murray, *Museums,* I, 39.

189. Lewis Carroll, *Through the Looking Glass,* in *Alice in Wonderland,* ed. Donald J. Gray (1871; New York: W. W. Norton, 1971), 175.

190. William Frederic Martyn, *A New Dictionary of Natural History; or Compleat Universal Display of Animated Nature* (London: Harrison, 1785), II, not paginated; for an overview of unicorn literature, see Odell Shepard, *The Lore of the Unicorn* (London: George Allen and Unwin, 1930), chap. 6.

191. In, for example, *Numbers* 24:8, *Deuteronomy* 33:17, *Isaiah* 34:7, *Job* 39:9–12, and *Psalms* 29:6. John Worcester, *Correspondences of the Bible: The Animals* (Boston: Massachusetts New-church Union, 1884), 30–31; Lulu Rumsey Wiley, *Bible Animals* (New York: Vantage, 1957), 428–431.

192. *Zoological Keepsake; or Zoology, and the Garden and Museum of the Zoological Society, for the year 1830* (London: Marsh and Miller, 1830), 69–70; William Jardine, *The Natural History of the Ruminating Animals, Part I. Containing Deer, Antelopes, Camels, &c* (Edinburgh: W. H. Lizars, 1835), 201.

193. Charles Gould, *Mythical Monsters* (London: W. H. Allen, 1886), 364.

194. John Barrow, *An Account of Travels into the Interior of Southern Africa, in the Years 1797 and 1798* (London: T. Cadell Jun. and W. Davies, 1801), I, 312–318.

195. *Curiosities for the Ingenious. Selected from the Most Authentic Treasures of Nature, Science and Art* (London: Thomas Boys, 1822), 107–108.

196. John Timbs, *Eccentricities of the Animal Creation* (London: Seeley, Jackson and Halliday, 1869), 58–59.

197. G. R. Blanche in "Notes," *Nature* 6 (1872), 292.

198. Timbs, *Eccentricities of the Animal Creation,* 53. For an anthology of Victorian

unicorn sightings, see Philip Henry Gosse, *The Romance of Natural History* (Boston: Gould and Lincoln, 1861), 285–290.

199. Gould, *Mythical Monsters,* 364.

200. William Clift, "Some account of an Object now exhibiting (November 1822) at the Turf Coffee-house . . . under the appellation of '*Mermaid*,'" manuscript, Royal College of Surgeons; Everard Home to William Clift, September 20, 1822, Royal College of Surgeons; *Morning Herald,* November 21, 1822.

201. "The Mermaid," *The Mirror of Literature, Amusement, and Instruction,* November 9, 1822, 17–18.

202. Thompson, *Mystery and Lore of Monsters,* 63; Richard Carrington, *Mermaids and Mastodons: A Book of Natural and Unnatural History* (London: Chatto and Windus, 1957), 14. For accounts of mermaids shown in London during the eighteenth and nineteenth centuries, see "The Mermaid," *Mirror,* November 9, 1822; Altick, *Shows of London,* 302–303; Peter Dance, *Animal Fakes and Frauds* (Maidenhead, Berks.: Sampson Low, 1976), chap. 4; and Carrington, *Mermaids and Mastodons,* ch. 1.

203. "To Be Seen Here, The Wonderful and Surprising Mermaid . . . ," poster in "Printed Ephemera" Box, Linnean Society Library; *Some Account of the Merman, Now Exhibiting at 174, Picadilly . . .* (London: W. Molineux, ca. 1826), 2, 16.

204. Major General Thomas Hardwicke, "Notes on Zoology" (1820s), I, 141, British Museum Add. Mss. 9879–80.

205. J. Murray, "Mermaid," *Magazine of Natural History* 3 (1830), 447.

206. "Mermaid," *Annals of Sporting and Fancy Gazette* 6 (1824), 126.

207. Carrington, *Mermaids and Mastodons,* 14.

208. Phil Robinson claimed that folk belief in mermaids and other mythological water creatures was still widespread among inhabitants of the entire British and Irish coasts, especially in areas that were or had been Celtic, in *Fishes of Fancy: Their Place in Myth, Fable, Fairy Tale and Folk-Lore; etc.* (London: William Clowes, 1883), 80.

209. *Naturalist's Pocket Magazine; or, Compleat Cabinet of the Curiosities and Beauties of Nature* (London: Harrison, Cluse, 1799), vol. II, not paginated; Wernerian Natural History Society Minute Books, vol. I, 196, Edinburgh University Library Special Collections; Henry Lee, *Sea Fables Explained* (London: William Clowes, 1883), 37–41.

210. C. H. Mayo, "Mermaids," *Notes and Queries,* 6th ser., 5 (1882), 478; S. Barton-Eckett, "Eating a Mermaid," *Notes and Queries,* 5th ser., 3 (1875), 274.

211. St. George Mivart, *American Types of Animal Life* (Boston: Little, Brown, 1893), 311.

212. Martyn, *New Dictionary of Natural History,* II, not paginated; William MacLeay, "Draft Classification of Mammals on Quinary Principles," 1829, not paginated, Linnean Society archives.

213. William Swainson, *On the Natural History and Classification of Quadrupeds* (London: Longman, Rees, Orme, Brown, Green and Longman, 1835), 97.

214. Gould, *Mythical Monsters,* 2.

215. "The Giraffe," *Zoological Magazine, or Journal of Natural History,* no. 1 (1833), 1.

216. Gosse, *Romance of Natural History,* 296.

217. One of the conventions of sea serpent scholarship, both in the nineteenth century and subsequently, has been the detailed chronological recounting, often with extensive quotation from contemporary journalistic and scientific sources, of sightings. See, for example, Gosse, *Romance of Natural History,* chap. 12; Gould, *Mythical Monsters,* 289–303; A. C. Oudemans, *The Great Sea-Serpent. An Historical and Critical Treatise* (Leiden: E. J. Brill; London: Luzac, 1892); Bernard Heuvelmans, *In the Wake of the Sea-Serpents,* trans. Richard Garnett (New York: Hill and Wang, 1968); Carrington, *Mermaids and Mastodons,* chap. 2; Gary Mangiacopra, "The Great Unknowns of the Nineteenth Century," *Of Sea and Shore* (Winter 1976–77), 201–205, 228; Charles Bright, *Sea Serpents* (Bowling Green, Ohio: Bowling Green State University Popular Press, 1991); and Richard Ellis, *Monsters of the Sea* (New York: Alfred A. Knopf, 1994), 39–74.

218. *Nature* 6 (1872), 402.

219. "The Great Sea Serpent," *Daily News,* November 16, 1848.

220. "The Great Sea-Serpent," *Illustrated London News* (October 28, 1848), 264–266.

221. "The Great Sea Serpent as Seen from the Ship Hydaspes," *Police News,* May 20, 1876.

222. Ms. letter from P[atrick] Neill to C. Lyell, March 14, 1848, in "Notes and press-cuttings about Sea Serpents," Edinburgh University Library Special Collections, Lyell 5.

223. Henry Lee, *Sea Monsters Unmasked* (London: William Clowes and Sons, 1883), 103.

224. T. H. Perkins to his father, October 13, 1820. Ms. copy in Charles Lyell, "Notes and press-cuttings about sea serpents," Edinburgh University Library Special Collections. For an extended consideration of the issues of Yankee and American credibility at stake in this controversy, see Chandos Michael Brown, "A Natural History of the Gloucester Sea Serpent: Knowledge, Power, and the Culture of Science in Antebellum America," *American Quarterly* 42 (1990), 402–436.

225. Andrew Wilson, *Facts and Fictions of Zoology* (New York: J. Fitzgerald, 1882), [218], 12; Ron Westrum, "Knowledge about Sea-Serpents," in Roy Wallis, ed., *On the Margins of Science: The Social Construction of Rejected Knowledge* (Keele: University of Keele, 1979), 298–300.

226. Heuvelmans, *In the Wake of the Sea-Serpents,* 274–276; letters and manuscript notes in Richard Owen's scrapbook, "Sea Serpentines," British Museum (Natural History) Archives; Frank Buckland, "Practical Natural History: 'The Great Sea Serpent,'" *Land and Water* (September 8, 1877), 196–198.

227. Robert Hamilton, *The Natural History of Amphibious Carnivora, including the Walrus and Seals, also of the Herbivorous Cetacea* (Edinburgh: W. H. Lizars, 1839), 326.

228. Buckland, "Practical Natural History: 'The Great Sea Serpent,'" 196.

229. Richard Owen, "Sea Serpentines" Scrapbook, British Museum (Natural History) Archives.

230. Nicolaas Rupke has suggested that Owen's anti-sea-serpent rhetoric was specifically

directed against amateur naturalists who preferred legalistic to scientific modes of argumentation. *Richard Owen: Victorian Naturalist* (Chicago: University of Chicago Press, 1994), 324–332.

5. Matters of Taste

1. Sawrey Gilpin, "On the character and expression of Animals," 5, 13, Bodleian Ms. Eng. misc.d.585 (folio 5–21), Bodleian Library, Oxford.

2. Benjamin R. Haydon, *Lectures on Painting and Design* (London: Longman, Brown, Green, and Longmans, 1844–1846), I, 13.

3. Joseph Strutt, *The Sports and Pastimes of the People of England,* ed. William Hone (London: Chatto and Windus, 1876), 75.

4. Rowland Ward, *The Sportsman's Handbook to Practical Collecting, Preserving, and Artistic Setting-Up of Trophies and Specimens* (London: Simpkin, Marshall, Hamilton, Kent, 1890), 7.

5. Quoted from William Ellis, *The Modern Husbandman* (1750), in James Britten, *Old Country and Farming Words: Gleaned from Agricultural Books* (London: Trübner, 1880), 43; Duncan George Forbes MacDonald, *Cattle, Sheep, and Deer* (London: Steele and Jones, 1872), 574–575.

6. P. B. Munsche, *Gentlemen and Poachers: The English Game Laws, 1671–1831* (Cambridge: Cambridge University Press, 1981), 3–4; "Game Laws and Game Preserving," *Westminster Review* 114 (1880), 134–135.

7. "Game Laws and Game Preserving," 144.

8. "Forest and Game Laws," *Westminster Review* 45 (1846), 407, 417.

9. [J. S. Mill], "The Game Laws," *Westminster Review* 5 (1826), 3.

10. Munsche, *Gentlemen and Poachers,* 23–25, 143.

11. [Mill], "Game Laws," 15.

12. George Graves, *The Naturalist's Companion, Being a Brief Introduction to the Different Branches of Natural History* (London: Longman, Hurst, Rees, Orme, Brown, and Green, 1824), 44.

13. James Buckman, "Vermin," in John Morton, ed., *A Cyclopedia of Agriculture, Practical and Scientific* (Glasgow: Blackie and Son, 1855), II, 1060.

14. John Donaldson, *The Enemies to Agriculture, Botanical and Zoological* (London: Robert Baldwin, 1847), 71.

15. Buckman, "Vermin," II, 1060.

16. Robert Smith, *The Universal Directory for Taking Alive and Destroying Rats, and all other kinds of four-footed and winged vermin, in a manner hitherto unattempted: calculated for the use of the gentleman, the farmer, and the warrener* (London: the Author, 1768).

17. See, for example, Buckman, "Vermin," II, 1061–1064.

18. Donaldson, *Enemies to Agriculture,* 79, 80.

19. [Mill], "Game Laws," 13.

20. Buckman, "Vermin," II, 1063; Donaldson, *Enemies to Agriculture,* 75.

21. Georges Louis Leclerc, Comte de Buffon, *Barr's Buffon. Buffon's Natural History* (London: H. D. Symonds, 1797), VI, 115; William MacGillivray, *A History of British Quadrupeds* (Edinburgh: W. H. Lizars, 1838), 62.

22. For a discussion of the ambiguous legal position of eighteenth-century dogs, see Susan Staves, "The Reasonable Dog and the Reasonable Dog Owner in the Age of Enlightenment," paper delivered at American Society for Eighteenth-Century Studies (Providence, 1993).

23. Smith, *Universal Directory*, 47.

24. Francis T. Buckland, *Curiosities of Natural History. Second Series* (1860; rpt., London: Macmillan, 1900), 72–73.

25. Harriet Ritvo, *The Animal Estate: The English and Other Creatures in the Victorian Age* (Cambridge, Mass.: Harvard University Press, 1987), 88–89.

26. Mary Campbell Smith, "Unprotected Vermin," *Westminster Review* 143 (1895), 204–205.

27. The symbolic significance of meat eating, as of other patterns of food consumption and avoidance, has been analyzed by anthropologists and sociologists concerned with a range of human societies. See, for example, Mary Douglas, *Purity and Danger: An Analysis of Concepts of Pollution and Taboo* (London: Routledge, 1966), and "Deciphering a Meal," in *Implicit Meanings: Essays in Anthropology* (London: Routledge and Kegan Paul, 1975), 249–275; Marcel Détienne and Jean-Pierre Vernant, eds., *The Cuisine of Sacrifice among the Greeks*, trans. Paula Wissing (Chicago: University of Chicago Press, 1989); Jack Goody, *Cooking, Cuisine and Class: A Study in Comparative Sociology* (Cambridge: Cambridge University Press, 1982); Julia Twigg, "Vegetarianism and the Meanings of Meat," in Anne Marcott, ed., *The Sociology of Food and Eating: Essays on the Sociological Significance of Food* (Aldershot, Hants.: Gower, 1983), 18–30; and Nick Fiddes, *Meat: A Natural Symbol* (London: Routledge, 1991).

28. John Arbuthnot, *An Essay Concerning the Nature of Aliments, and the Choice of Them, According to the Different Constitutions of Human Bodies* (London: J. Tonson, 1732), 225; J. Milner Fothergill, *The Food We Eat: Why We Eat It and Whence It Comes* (London: Griffith and Farran, 1882), 53–54.

29. Tabitha Tickletooth [Charles Selby], *The Dinner Question: or, How to Dine Well and Economically* (London: Routledge, Warne, and Routledge, 1860), 37.

30. C. Anne Wilson, *Food and Drink in Britain, from the Stone Age to Recent Times* (London: Constable, 1973), 97; Frederick William Hackwood, *Good Cheer: The Romance of Food and Feasting* (New York: Sturgis and Walton, 1911), 172, 304.

31. John Burnett, *Plenty and Want: A Social History of Diet in England from 1815 to the Present Day* (London: Methuen, 1983), 64, 88; *Hints for the Table: or the Economy of Good Living* (London: George Routledge and Sons, 1866), 58.

32. Eleanor E. Orlebar, *Food for the People; or, Lentils and Other Vegetable Cookery* (London: Sampson, Low, Marston, Searle and Rivington, 1879), 16.

33. William Cobbett, *Cottage Economy* (London: C. Clement, 1822), 162. During the early nineteenth century, many of the rural and urban poor went largely without

meat, although well-paid industrial workers could eat meat every day. Betty McNamee, "Trends in Meat Consumption," in T. C. Barker, J. C. McKenzie, and John Yudkin, eds., *Our Changing Fare: Two Hundred Years of British Food Habits* (London: MacGibbon and Kee, 1966), 76–79.

34. J. S. Forsyth, *The Natural and Medical Dieteticon: or, Practical Rules for Eating, Drinking, and Preserving Health, on Principles of Easy Digestion* (London: Sherwood, Jones, 1824), 236.

35. "On the Origin and Natural History of the Domestic Ox, and its Allied Species," *Quarterly Journal of Agriculture* 2 (1830), 196–197.

36. Quoted in Peter Lund Simmonds, *The Curiosities of Food; or the Dainties and Delicacies of Different Nations Obtained from the Animal Kingdom* (London: R. Bentley, 1859), 2–3.

37. Anne Walbank Buckland, *Our Viands: Whence They Come and How They Are Cooked, with a Bundle of Old Recipes from Cookery Books of the Last Century* (London: Ward and Downey, 1893), 75; Arthur Robert Kenney-Herbert, *Culinary Jottings for Madras, or, A Treatise in Thirty Chapters on Reformed Cookery for Anglo-Indian Exiles*, ed. Leslie Forbes (1885; rpt., Prospect Books: Totnes, Devon, 1994), 102–103.

38. William Bingley, *Useful Knowledge; or a Familiar Account of the Various Productions of Nature, Mineral, Vegetable, and Animal, which are Chiefly Employed for the Use of Man* (London: Baldwin and Craddock, 1831), III, 94; MacDonald, *Cattle, Sheep, and Deer*, 8; Duncan McDonald, *The New London Family Cook; or, Town and Country Housekeeper's Guide* (London: Albion Press, 1808), 25.

39. *The Guide to Service: The Cook* (London: Charles Knight, 1842), 12–13.

40. Antonio Celestino Cocchi, *The Pythagorean Diet, of Vegetables Only, Conducive to the Preservation of Health and the Cure of Diseases* (London: R. Dodsley, 1745), 28–29, 69.

41. John Frank Newton, *The Return to Nature* (1811; London: Ideal Publishing Union, 1897), 74.

42. William H. Pyne, *Microcosm: or, A Picturesque Delineation of the Arts, Agriculture, Manufactures, &c. of Great Britain* (London: W. H. Pyne and J. C. Nattes, 1806), I, 13.

43. Joseph Ritson, *An Essay on Abstinence from Animal Food as a Moral Duty* (London: Richard Phillips, 1802), quoted in Frederick J. Simoons, *Eat Not This Flesh: Food Avoidances in the Old World* (Madison: University of Wisconsin Press, 1961), 11.

44. *Pure Food for All People. Plenty, Nutritious, Tasty* (London: United Templar's Society, n.d. [late 1800s]), not paginated.

45. For an argument based on comparative anatomy, see Anna Kingsford, "The Best Food for Man," *Westminster Review* 102 (1874), 500–514.

46. See Molly Baer Kramer, "A Resurgence of Compassion: Three Animal Protection Movements in England, 1950–1980," d.Phil. Diss., Oxford, 1996; Elizabeth Driver, *A Bibliography of Cookery Books Published in Britain, 1875–1914* (London: Prospect Books, 1989), 27.

47. Orlebar, *Food for the People*, 77–80.

48. Charles W. Forward, *Practical Vegetarian Recipes as Used in the Principal Vegetarian Restaurants in London and the Provinces* (London: J. S. Virtue, 1891), 6, 13.

49. *Punch* 80 (May 28, 1881), 251.

50. Burnett, *Plenty and Want*, 228.

51. Forward, *Practical Vegetarian Recipes*, 7.

52. Joseph Haslewood, *Some Account of the Life and Publications of the Late Joseph Ritson, Esq.* (London: Robert Triphook, 1824), 32.

53. Quoted in George Hendrick, *Henry Salt: Humanitarian Reformer and Man of Letters* (Urbana, Ill.: University of Illinois Press, 1977), 20.

54. J. A. Paris, *A Treatise on Diet: With a View to Establish, on Practical Grounds, A System of Rules for the Prevention and Cure of the Diseases Incident to a Disordered State of the Digestive Functions* (London: Sherwood, Gilbert and Piper, 1837), 128.

55. R. Govett, *Vegetarianism: A Dialogue* (Norwich: Fletcher, 1883), 13.

56. Edmund S. and Ellen J. Delamere, *Wholesome Fare or the Doctor and the Cook: A Manual of the Laws of Food and the Practice of Cookery* (London: Lockwood, 1868), 3, 341–342.

57. Forsyth, *Natural and Medical Dieteticon*, 75.

58. Anna K. Eccles, *A Manual of What to Eat and How to Cook It for Salisbury Patients* (New York: Kellogg, 1897), 15, 63–64.

59. "Feeding the Negro White," *Lancet* (June 2, 1860), 555.

60. Buckland, *Our Viands*, 75.

61. Thomas Moffett, *Health's Improvement: Or, Rules Comprizing and Discovering the Nature, Method and Manner of Preparing All Sorts of Foods Used in this Nation* (London: T. Osborne, 1746), 103.

62. *Hints for the Table: or, the Economy of Good Living* (London: George Routledge and Sons, 1866), 7.

63. See, for example, *Guide to Service: The Cook*, 38.

64. Wilson, *Food and Drink in Britain*, 109.

65. Paris, *Treatise on Diet*, 187–188.

66. Mary Barrett Brown, *Fish, Flesh and Fowl: When in Season, How to Select, Cook, and Serve* (London: L. Upcott Gill, 1897), 46; R. W. Dickson, *A Complete System of Improved Live Stock and Cattle Management* (London: Thomas Kelly, 1824), I, 28.

67. Robert Mudie, *Domesticated Animals, Popularly Considered, in their Structure, Habits, Localities, Distribution, Natural Relations, and Influence upon the Progress of Human Society* (Winchester: D. E. Gilmour, 1839), 241–242; Moffett, *Health's Improvement*, 126.

68. Brown, *Fish, Flesh and Fowl*, 46; *Guide to Service: The Cook*, 229

69. Thomas Bewick, *A General History of Quadrupeds* (Newcastle-upon-Tyne: T. Bewick and Son, 1824), 512.

70. William Ewart, "Meat," in John C. Morton, ed., *A Cyclopedia of Agriculture, Practical and Scientific* (Glasgow: Blackie and Son, 1855), II, 396.

71. Thomas Webster and Mrs. Parkes, *An Encyclopedia of Domestic Economy* (London: Longman, Brown, Green, and Longman, 1847), 375.

72. Webster and Parkes, *Encyclopedia of Domestic Economy,* 377, 379.

73. *Guide to Service: The Cook,* 230.

74. See Harriet Ritvo, "Race, Breed, and Myths of Origin: Chillingham Cattle as Ancient Britons," *Representations* 39 (1992), 1–22.

75. Frank Graham, *Wooler, Ford, Chillingham and the Cheviots* (Newcastle: Frank Graham, 1976), 30; Charles Oldham, "The Lyme Park Herd of Wild White Cattle," *Zoologist* 15 (1891), 85.

76. Charles G. Barrett, "The Wild Cattle at Chillingham," *Transactions of the Norfolk and Norwich Naturalists Society* (1875), 54; Laisters F. Lort, "The White Cattle of Vaynol Park," *Transactions of the North Staffordshire Field Club* 33 (1898–1899), 56.

77. Jack C. Drummond and Anne Wilbraham, *The Englishman's Food: A History of Five Centuries of English Diet* (London: Jonathan Cape, 1939), 119.

78. Maxine McKendry, *Seven Hundred Years of English Cooking,* ed. Arabella Boxer (London: Treasure Press, 1973) 14; Wilson, *Food and Drink in Britain,* 83–84.

79. Bingley, *Useful Knowledge,* 48.

80. Robert Hamilton, *The Natural History of the Ordinary Cetacea or Whales* (Edinburgh: W. H. Lizars, 1837), 243.

81. John Goodsir, "Lectures on Comparative Anatomy," student notes, Summer Session, 1858 (Lecture 2, May 20th), Edinburgh University Library Special Collections, Gen 580D–581D.

82. Moffett, *Health's Improvement,* 158.

83. Vincent M. Holt, *Why Not Eat Insects?* (1885; rpt., London: British Museum (Natural History), 1988), 3, 4, 9–10.

84. Holt, *Why Not Eat Insects?,* 96–99.

85. Holt, *Why Not Eat Insects?,* 24, 26.

86. Simoons, *Eat Not This Flesh,* 85.

87. W. W. Cazalet, "Hippophagy and Onophagy," *Temple Bar* 19 (1866), 31–32; Jean-Pierre Digard, *L'homme et les animaux domestiques: Anthropologie d'une passion* (Paris: Fayard, 1990), 72–73.

88. Simoons, *Eat Not This Flesh,* 84; quoted in Wilson, *Food and Drink in Britain,* 76.

89. Maguelonne Toussaint-Samat, *The History of Food* (Cambridge, Mass.: Blackwell, 1992), 98.

90. Simoons, *Eat Not This Flesh,* 85; Cazalet, "Hippophagy and Onophagy," 33–34.

91. "Horse-flesh as Food for Man," *Lancet* (October 31, 1857), 457.

92. Drummond and Wilbraham, *The Englishman's Food,* 364–365.

93. Cazalet, "Hippophagy and Onophagy," 34; "Horse-flesh as Food for Man," 457.

94. An Old Bohemian, *Dishes and Drinks; or, Philosophy in the Kitchen* (London: Ward and Downey, 1887), 52–53.

95. William Burchell, *Travels in the Interior of Southern Africa* (London: Longman, Hurst, Rees, Orme, and Brown, 1822, 1824), II, 83, 238.

96. See Goody, *Cooking, Cuisine and Class,* 146.

97. Quoted in Adrian Desmond, *Huxley: The Devil's Disciple* (London: Michael Joseph, 1994), 58.

98. Thomas Hardwicke, "Description of a Species of Jerboa, found in the upper Provinces of Hindustan . . . ," *Linnean Society of London Transactions* 8 (1804), 281.

99. Bingley, *Useful Knowledge*, III, 53–54, 40.

100. Peter Lund Simmonds, *Animal Products: Their Preparation, Commercial Uses, and Value* (New York: Scribner, Welford, and Armstrong: 1877), 1, 2.

101. Quoted in J. R. H. Andrews, *The Southern Ark: Zoological Discovery in New Zealand, 1769–1900* (Honolulu: University of Hawaii Press, 1986), 9.

102. G. H. O. Burgess, *The Eccentric Ark: The Curious World of Frank Buckland* (New York: Horizon Press, 1967), 11; George C. Bompas, *Life of Frank Buckland* (London: Smith, Elder, 1885), 46, 128.

103. Janet Browne, *Charles Darwin: Voyaging* (New York: Knopf, 1995), 107.

104. Ian Cameron, *The History of the Royal Geographical Society, 1830–1980: To the Farthest Ends of the Earth* (London: Macdonald and Jane's, 1980), 16; Charles Davies Sherborn, comp., *The Zoological Club, 1866–1932* (London: Zoological Club, 1933), 3–5.

105. Webster and Parkes, *Encyclopedia of Domestic Economy*, 380.

106. A Beef Eater, *Illustrations of Eating: Displaying the Omnivorous Character of Man; and Exhibiting the Natives of Various Countries at Feeding Time* (London: John Russell Smith, 1847), 36–37.

107. W. Taplin, *The Sportsman's Cabinet* (London, 1803), I, 14.

108. Bingley, *Useful Knowledge*, III, 24–25, 38.

109. Hugh Dalziel, *British Dogs: Their Varieties, History, Characteristics, Breeding, Management, and Exhibition* (London: "The Bazaar," 1881), 447.

110. An Old Bohemian, *Dishes and Drinks*, 54.

111. Holt, *Why Not Eat Insects?*, 11.

112. Webster and Parkes, *Encyclopedia of Domestic Economy*, 371.

113. Cazalet, "Hippophagy and Onophagy," 35; Frederick Markham, *Shooting in the Himalayas. A Journal of Sporting Adventures and Travel in Chinese Tartary, Ladac, Thibet, Cashmere* (London: Richard Bentley, 1854), 145.

114. Simoons, *Eat Not This Flesh*, 104.

115. Webster and Parkes, *Encyclopedia of Domestic Economy*, 380.

116. Charles Darwin, *The Variation of Animals and Plants under Domestication* (New York: D. Appleton, 1892), II, 205.

117. A Beef Eater, *Illustrations of Eating*, 38.

118. Hackwood, *Good Cheer*, 290.

119. *The British Chronicle*, November 14, 1771, quoted in Clifford Morsley, *News from the English Countryside, 1750–1850* (London: Harrap, 1979), 71–72.

120. Gordon Stables, *Dogs in Their Relation to the Public (Social, Sanitary and Legal)* (London: Cassell, Petter, and Galpin, 1877), 14–15.

121. "Eating Cats at West Bromwich," *Live Stock Journal and Fancier's Gazette* 2 (1875) 756.

122. Thomas Williams, quoted in George Stocking, *Victorian Anthropology* (New York: Free Press, 1987), 90.

123. Buckland, *Our Viands,* 9.

124. W. Linnaeus Martin, *A General Introduction to the Natural History of Mammiferous Animals, with a Particular View of the Physical History of Man* (London: Wright, 1841), 267.

125. British Association for the Advancement of Science, *Notes and Queries on Anthropology for the Use of Travellers and Residents in Uncivilized Lands* (London: Edward Stanford, 1874), 45.

126. *Punch* 68 (January 16, 1875), 31.

127. See, for example, *Nature* 35 (1887), 350 (untitled note).

128. "Scientific Serials," *Nature* 8 (1873), 375.

129. "London Phrenological Society," *Lancet* 1 (1832), 488.

130. For example, the Monbuttas "ranked as one of the most important monarchical states of Central Africa . . . [and] recognised one supreme being." The Battas of Sumatra were "so far advanced in civilization" that they "actually have laws to regulate the eating of criminals and prisoners of war"; according to Stamford Raffles, they too "acknowledged a Supreme Being." "Scientific Serials," 375; A Beef Eater, *Illustrations of Eating,* 40; Hackwood, *Good Cheer,* 339.

131. See Anthony Pagden, *The Fall of Natural Man: The American Indian and the Origins of Comparative Ethnology* (Cambridge: Cambridge University Press, 1982), 81–86, for a discussion of such claims with regard to the indigenous inhabitants of the Caribbean.

132. Thomas Henry Huxley, *Man's Place in Nature* (1863; rpt., Ann Arbor: University of Michigan Press, 1959), 69–70.

133. Patrick Brantlinger, *Rule of Darkness: British Literature and Imperialism, 1830–1914* (Ithaca, N.Y.: Cornell University Press, 1988), 185–186.

134. For instances of eighteenth- and nineteenth-century maritime cannibalism, see A. W. Brian Simpson, *Cannibalism and the Common Law: The Story of the Tragic Last Voyage of the Mignonette and the Strange Legal Proceedings to Which It Gave Rise* (Chicago: University of Chicago Press, 1984), 114–143.

135. See Simpson, *Cannibalism and the Common Law,* 147–160, for land expeditions similarly driven to cannibalism.

136. *The Discovery of Sir John Franklin and Thirty of his Crew* (Bristol: John Chapman, n.d. [1854]), not paginated.

137. Simpson, *Cannibalism and the Common Law,* 10–11, 89, 248–250.

138. "Cannibalism," *Lancet* (September 21, 1867), 373.

139. [Brougham], "Ritson on Abstinence from Vegetable Food," *Edinburgh Review* 2 (1803), 133.

140. Samuel Phillips, *Guide to the Crystal Palace and Park* (London: Crystal Palace Library, 1856), 118.

141. Stocking, *Victorian Anthropology,* 252.

Acknowledgments

M ANY PEOPLE have helped me as I have worked on this book. I am especially indebted to friends and colleagues who read the whole manuscript: Rosemarie Bodenheimer, Ilona Karmel, and Andy Von Hendy in installments; and Janet Browne, Peg Fulton, Marie-Hélène Huet, and Jim Secord when it was more or less finished. I am also grateful to Tom Metcalf, Craig Murphy, Dorinda Outram, Jim Paradis, and JoAnne Yates for their advice on particular sections. The Undergraduate Research Opportunities Program at MIT provided me with a series of enthusiastic and able research assistants: Mary Pat Reeve, Jessica Marcus, Sherrian Lea, Teresa Esser, and Jennifer Lee.

I appreciate the assistance offered by the librarians and archivists at the following institutions: the Massachusetts Institute of Technology; Widener Library, Houghton Library, the Ernst Mayr Library, Schlesinger Library, and Countway Library, at Harvard University; the University Library and the Museum of Zoology at Cambridge University; the Natural History Museum (UK); the Edinburgh University Library; the Glasgow University Library; the Kennel Club; the Linnean Society; the Royal College of Surgeons; the Bodleian Library, the Hope Library, the University Museum, and the Ashmolean Museum at Oxford University; the British Library; the Wellcome Institute for the History of Medicine; the Yale Center for British Art; the Library of Congress; the National Agricultural Library (US); the Boston Public Library; the Horniman Museum; the Saffron Walden Museum; the National Art Library (UK); the Bristol, Clifton, and West of England Zoological Society; and the county record offices of Essex, Suffolk, Cambridgeshire, and Gloucestershire. The Ashmolean Museum; the Bodleian Library;

the Bristol, Clifton, and West of England Zoological Society; the British Library; the Cambridge University Library; the Cambridge University Museum of Zoology; the Edinburgh University Library; the Glasgow University Library; the Linnean Society; the Natural History Museum; and the Royal College of Surgeons have kindly allowed me to quote from manuscript material in their possession.

Fellowships from the Guggenheim Foundation, the National Endowment for the Humanities, and the National Humanities Center gave me time for research and writing. I am also grateful for support of various kinds from the Whiting Foundation; the School of Humanities and Social Science at the Massachusetts Institute of Technology; Clare Hall, Cambridge; and Balliol College, Oxford.

My research has required that I spend long periods in the United Kingdom. This necessity has always been a pleasure, and I have been fortunate in the hospitality of individuals as well as institutions. I would like especially to thank my friends Andrew, Jennifer, Henry, George, and Eleanor Warren, who have made me feel like a member of their family.

Illustration Credits

Index